Lecture Notes on Multidisciplinary Industrial Engineering

Series editor

J. Paulo Davim, Department of Mechanical Engineering, University of Aveiro, Aveiro, Portugal

"Lecture Notes on Multidisciplinary Industrial Engineering" publishes special volumes of conferences, workshops and symposia in interdisciplinary topics of interest. Disciplines such as materials science, nanosciences, sustainability science, management sciences, computational sciences, mechanical engineering, industrial engineering, manufacturing, mechatronics, electrical engineering, environmental and civil engineering, chemical engineering, systems engineering and biomedical engineering are covered. Selected and peer-reviewed papers from events in these fields can be considered for publication in this series.

More information about this series at http://www.springer.com/series/15734

Uday Shanker Dixit · R. Ganesh Narayanan
Editors

Strengthening and Joining by Plastic Deformation

Select Papers from AIMTDR 2016

 Springer

Chapters 3–10 concentrate on joining of materials involving plastic deformation. Electromagnetic pulse crimping of aluminum tube on dual phase steel rod is attempted in Chap. 3. Electromagnetic pulse joining is a sustainable solid-state joining method. Hardness and pull-out tests were conducted to ascertain the mechanical strength of the joint. Chapter 4 describes the post-weld heat treatment processes on the corrosion and mechanical properties of friction stir welded 7xxx aluminum alloy. Friction stir spot welding of polymer core sandwich sheets is discussed in Chap. 5. The temperature evolution, hardness distribution, joint formation and static strength of the joints were evaluated to reveal the validity of the process for joining sandwich sheets. Chapter 6 focuses on the friction stir welding of thermoplastic polymers. A case study on welding of high-density polyethylene polymer aiming at studying the effect of welding speed and rotational speed on the joint mechanical properties was performed. Chapter 7 presents an enhancement of mechanical properties of friction stir welded portion of Al 6061 by incorporating additional reinforcing particulates of silicon carbide and aluminum oxide at the weld interface. Ultrasonic spot welding of aluminum and copper sheets is discussed in Chap. 8. Mechanical and metallurgical tests confirmed the application of the welding method for joining of dissimilar metal sheets. Finite element simulations were also conducted to study the deformation and stress distribution during joining. Chapter 9 describes the finite element simulation of magnetic pulse welding for predicting the distribution of electromagnetic force and magnetic field. Chapter 10 discusses the potential of electromagnetic welding for fabricating tubular structures in nuclear power industry. Joining of grades of aluminum alloys and stainless steel was experimented, and interesting results have been presented.

These chapters highlight the significance of strengthening and joining by plastic deformation for a variety of materials including metals and polymers. Future directions for research are also provided. It is expected that the book will be welcomed by engineers and scientists, particularly those involved in mechanical, metallurgical, material science, manufacturing and industrial engineering fields. We solicit the valuable feedback of the readers.

Guwahati, India   Uday Shanker Dixit
                                                                        R. Ganesh Narayanan

Editorial Acknowledgements

We are thankful to Professor B. B. Ahuja, Director of College of Engineering Pune (COEP), who was chairman and organizing secretary of AIMTDR 2016, for allowing us to publish the book from extending versions of selected papers of the conference. Thanks to Dr. B. Rajiv, Head of Department of Production Engineering and Industrial Management, College of Engineering Pune (COEP), and the organizing team for their motivation and encouragement provided during conference organization that led to the publication of this book. We thank all the authors for contributing their valued research work as extended chapters for this book. The authors submitted the chapters and responded to the revisions promptly which could make it possible to publish the book without delay. We sincerely acknowledge the organizers and the National Advisory Committee (NAC) members of AIMTDR 2016 for their guidance and support. We are grateful to Prof. Amitabha Ghosh, Prof. V. K. Jain, Prof. V. Radhakrishnan, Prof. M. S. Shunmugam and Prof. P. Radhakrishnan for their encouragement and suggestions.

We express our heartfelt appreciation to Prof. J. Paulo Davim from University of Aveiro, the Series Editor of this book, for his encouragement and guidance. Last but not least, we express our sincere gratitude to the staff members of Springer, particularly to Ms. Swati Meherishi, Ms. Aparajita Singh, Dr. Mayra Castro and Ms. Christy, for their dedicated support during publication process of this book.

About the AIMTDR Conference

AIMTDR conference is a highly prestigious, biennial event organized in the field of mechanical and production engineering in India. The conference has a glorious history of organization since its inception. The first conference entitled 'All India Machine Tool Design and Research Conference' was held at Jadavpur University, Kolkata, in 1967. In early 90s, it was thought appropriate to widen the scope of the conference to encompass areas related to different manufacturing process technologies and systems. Accordingly, it was renamed as 'All India Manufacturing Technology, Design and Research' conference and the 16th in the series was organized at Central Machine Tool Institute, Bangalore, in 1994. It became an international event with the first international conference being held at Indian Institute of Technology Roorkee in 2006. The international conference aimed to bring together academicians, researchers and industry professionals working worldwide in the field of manufacturing to exchange and disseminate ideas.

The subsequent international AIMTDR conferences were held at IIT Madras and Andhra University, Visakhapatnam, respectively. Jadavpur University hosted silver jubilee of the conference and organized 4th International and 25th AIMTDR conference at Kolkata. The fifth international conference was organized at IIT Guwahati in December 2014, while the 6th one (AIMTDR 2016) was organized by Department of Production Engineering and Industrial Management, College of Engineering Pune (COEP), during December 16–18, 2016.

Mission, Vision, Challenges and Direction of AIMTDR Conference

(Excerpt from the address of Prof. Amitabha Ghosh, the Chief Guest of 26th AIMTDR conference, held at IIT Guwahati during December 12–14, 2014)

Personally, I have a close association with AIMTDR conference from the very first one for which I happened to be a humble and young member of the organizing team. That event was organized jointly by Late Prof. Amitabha Bhattacharyya and Prof. A. K. De at Jadavpur University in the year 1967. Keeping in view the growing industrialization of India, the AIMTDR conference was planned along the line of MTDR conference that used to be organized by Profs. Tobias and Koenigsberger at Cambridge and Birmingham alternately.

India being an emerging economy, the importance of 'manufacturing' was well recognized and one of the primary goals of AIMTDR conference has been to bring the academicians, researchers and the engineers from the industry to a common platform for exchanging ideas and developing a deeper mutual understanding among all concerned. The organizers of the AIMTDR conference in the past were eminently successful in this regard. With time, this event has gained maturity and has emerged as one of the most important international and national conferences held in India for all who are associated with the field of manufacturing.

Since the economic development of any country is very critically linked to the manufacturing sector, it is only very natural that the current political leadership of India has taken up 'Manufacturing in India' as one of its key objectives. In fact, 'manufacturing' should be a common objective for all the South Asian and the Southeast Asian countries to develop a good mutual understanding and cooperation to enhance the overall manufacturing capabilities of this region. Then only, this region of the world can become a powerhouse for economic growth and play a center-stage role in the world economy. This is essential if we have to eliminate the poverty that has plagued this region for a very long time.

From my long association with the evolution of AIMTDR conference and my involvement in teaching and R&D in the field of manufacturing for almost half a century, placing a few observations before this august gathering may not be out of place.

Traditionally, the 'primary' manufacturing processes have remained in the domain of mechanical engineering and metallurgy. At the same time, the secondary and finishing processes along with the machine tools and systems involved in manufacturing have remained exclusively as part and parcel of mechanical engineering. This, in my personal opinion, has not always helped the manufacturing activities in India to take advantage of the progress made in physical and applied sciences. Barring some isolated cases, this has rendered the manufacturing activities in India to be largely devoid of major fundamental innovations. As a result, the manufacturing activities in India have remained mostly confined to the traditional lines without giving much attention to 'value addition.' Thus, for example, our earnings from the export of a couple of hundred 'made in India' cars to Europe can be offset by that through the sale of a single focused ion beam machine, measuring $1.5 \text{ m} \times 1 \text{ m} \times 1 \text{ m}$, by USA to India. This scenario must be changed in the coming days.

Over the years, advanced manufacturing has gradually developed into a multi-disciplinary activity and real 'value addition' through 'manufacturing' can be achieved only when the advancement in physical, chemical and other sciences is used in innovating newer processes and possibilities of 'manufacturing.' Establishing a close link among the manufacturing and the related sciences has to be recognized as a necessary task. I believe that an event like AIMTDR can play a very significant role in this regard. Besides, revamping of the old-fashioned curricula for training engineers in manufacturing is essential to render them capable of facing the challenges from futuristic manufacturing; AIMTDR conference can take a leading role in that direction by providing a separate session to discuss the issues involved in the matter. In the coming years, I am very hopeful that AIMTDR conference will attract not only manufacturing engineers but also researchers from basic and applied sciences whose works are closely related to and important for innovations in manufacturing.

From my half-a-century-old teaching experience, I find that the current young generation is gradually becoming somewhat disinclined to take up careers in R&D related to manufacturing. This is particularly so for the brighter section of the student community; there is a feeling among them that there is not much intellectual challenge in the subjects related to manufacturing. There can be nothing farther from the truth. Perhaps, to a large extent, this is so as the curricula and syllabi have remained archaic in many universities and institutions. In fact, application of many advances in scientific principles to manufacturing is the key requirement to open the gate for the impending next Industrial Revolution. 'Manufacturing' needs the brilliant young minds to take up R&D careers in academia and industry with equal eagerness and enthusiasm. However, this can be possible only on receiving adequate and aggressive support from the industry houses—both financially and administratively.

Although the world has reaped the benefit of the 2nd Industrial Revolution which was triggered by R&D in the Silicon Valley, California, except for some software-related activities, India (and many other countries in this region) had really nothing to do with the actual developments and related manufacturing. Remaining a good follower cannot take India to any leadership position though it may provide some financial relief. India and the countries in this region cannot afford to miss the opportunity to take up important position when the next Industrial Revolution comes.

Dear colleagues and friends, the silicon-based 2nd Industrial Revolution has reached a plateau and there are indications that 'carbon' may play a more important role and carbon-based devices will play the key role in ushering the world into the reign of the 3rd Industrial Revolution. Already enough indications are coming in that direction. If that be the case, India should play a major role and initiate well-planned pioneering activities so that manufacturing engineering becomes a multidisciplinary area involving relevant basic science and engineering subjects for the emergence and growth of 'carbon-based technology' in this region of the world. I am happy to notice that some of the leading world authorities on carbon devices and advanced fabrication are present in this conference. This event can be a great opportunity for the manufacturing community to take advantage of their presence and plan an appropriate course of action to initiate planned activities to innovate carbon devices.

Many areas of manufacturing in the not-too-distant future will be very different from what we recognize as 'manufacturing' today; 'self-assembly,' 'self-regulation,' 'self-correction' and 'self-replication' will become the keywords in futuristic manufacturing. Obviously, it will be too drastic to think of redirecting all R&D on manufacturing in this direction, but, at the same time, India should be well prepared to take active role in such areas of futuristic manufacturing (a name for that was coined a few years ago in a workshop at IIT Kanpur—'Fabrionics') as that will help the country to gain expertise for incorporating significant 'value addition' in our manufacturing activities.

AIMTDR conference is one of the very few events that draw researchers and practitioners from the academia and the industry with equal enthusiasm. I am very hopeful that using this grand platform we all can take India and the neighboring countries along the path of growing technological excellence and engineering marvel.

AIMTDR 2016 Conference: Objectives and Organization

Manufacturing has revolutionized itself from its contemporary form to its current digital access, more so in the era of Industry 4.0. With every industrial revolution, we have seen labor and asset productivity multiply and structural shifts emerge in the manufacturing world order. Several core technologies are driving Industry 4.0, be it Simulation, Autonomous Robots, Big Data and Analytics, Augmented Reality or Additive Manufacturing. If our economy needs to grow multifold to achieve sustainable development, the existing 15% share of manufacturing sector in India's GDP needs to scale to 25% in the immediate near future.

Manufacturing today seeks innovation to be ubiquitous by inventing ways to produce more with less inputs. In an era of integration where technologies complement one another, design and manufacturing face a daunting task with regard to the quality and cost-effectiveness of products. Concentrated efforts focusing on quality research need to be endorsed for improving the manufacturing processes, technology and systems to adopt world-class manufacturing technologies. The manufacturing education should also emphasize its importance to attract the talented young to this area and equip them with skills that embrace knowledge, information and techniques.

With this broad focus, College of Engineering Pune (COEP) presented as a sequel to the AIMTDR conferences of the past, the 6th International and 27th AIMTDR conference in December 2016. The theme of the conference was 'Recontouring Manufacturing.' Several invited lectures and keynote addresses on cutting-edge technologies were presented in the conference by leading researchers from USA, Singapore and India. In all, 380 papers were selected for oral presentations and 80 for poster presentations. The organizing team brought out the proceedings on a CD covering all the papers presented. These papers will help to provide insights into the realistic exposure of current research and development trends in the field of Manufacturing Technology, Design and Research. During the exhibition in the conference, 40 leading companies participated by displaying new technology equipment, products and measuring equipment. The manufacturing community in India received this conference with appreciation and applause.

Through the deliberations in the conference, I hope to see the culmination of great thoughts and ideas that would introduce to develop technological solutions in the domain of manufacturing and design by the fusion of technologies straddling physical, digital and biological worlds.

B. B. Ahuja
Organizing Secretary, AIMTDR 2016
Professor of Production Engineering and Director
College of Engineering Pune

Contents

About the Editors

Dr. Uday Shanker Dixit is a Professor of Mechanical Engineering at Indian Institute of Technology Guwahati, India. His research interests include Metal Forming, Finite Element Method and Soft Computing-Based Modeling of Manufacturing Processes and Mechatronics. He has published more than 180 research papers in various journals and conferences and has authored/co-authored six books in Mechanical Engineering. He has also co-edited three books related to manufacturing. He is an Associate Editor of *Journal of Institution of Engineers (India) Series C*, the Regional Editor (Asia) of *International Journal of Mechatronics and Manufacturing Systems* and Dy. Section Editor of *Journal of Engineering*. Presently, he is the Vice President of National Advisory Committee of AIMTDR.

Dr. R. Ganesh Narayanan is an Associate Professor at Department of Mechanical Engineering, Indian Institute of Technology Guwahati, India. He received his Ph.D. from the Indian Institute of Technology Bombay, India. His research areas of interest include Metal Forming and Joining. He has contributed many research articles in reputed journals and international conferences and few edited books including Advances in Material Forming and Joining, Springer, India, and Metal Forming Technology and Process Modeling, McGraw Hill Education, India.

Chapter 1
Enhancement of Fatigue Life of Thick-Walled Cylinders Through Thermal Autofrettage Combined with Shrink-Fit

S. M. Kamal and Uday Shanker Dixit

Abstract Thick-walled cylindrical components are used in many industries, e.g. oil and chemical industries, artillery industries and nuclear power plants for withstanding high pressure or thermal gradient. Such components are subjected to autofrettage prior to their use in service, which increases their load carrying capacity as well as fatigue life. The fatigue life of the cylinder is important when the cylinder is subjected to a fluctuating or repeated pressure. Thermal autofrettage is a potential process capable of increasing the pressure carrying capacity as well as the thermal gradient capacity of thick-walled cylinders. This is achieved by employing a radial thermal gradient across the wall thickness of the cylinder. Due to the beneficial compressive residual stresses generated at and around the inner wall of the cylinder as a result of unloading of the thermal gradient, the thermally autofrettaged cylinder enhances the load carrying capacity as well as the fatigue life. Further enhancement in the fatigue life can be achieved by combining thermal autofrettage with shrink-fit. In this work, the fatigue life analysis of the thermally autofrettaged cylinder with shrink-fit is carried out. The analysis of thermal autofrettage is based on the assumptions of a generalized plane strain condition and Tresca yield criterion.

Keywords Thermal autofrettage · Thick-walled cylinder · Shrink-fit Stress intensity factor · Fatigue life · Paris law

S. M. Kamal (✉)
Department of Mechanical Engineering, School of Engineering, Tezpur University, Tezpur 784028, India
e-mail: smkmech@tezu.ernet.in

U. S. Dixit
Department of Mechanical Engineering, Indian Institute of Technology Guwahati, Guwahati 781039, India
e-mail: uday@iitg.ernet.in

1.1 Introduction

Autofrettage is a metal working process that induces compressive residual stresses at and around the inner wall of a thick-walled cylinder by means of partial plastic deformation at the inner side. In the most commonly practiced hydraulic autofrettage process, the plastic deformation at the inner side of the cylinder is achieved by applying an ultra-high hydraulic pressure [7]. Another autofrettage process practiced in industries is the swage autofrettage. In this process, the material at the inside region of the cylinder is deformed plastically by means of forcing an extra large mandrel along the bore of the cylinder [5]. Firing explosives at the inner side of the cylinder is another way of producing desired partial plastic deformation of the inner wall of the cylinder. This method is known as explosive autofrettage [15]. When the loads causing the partial plastic deformation in the cylinder are released, beneficial compressive residual stresses are set up at the inside of the cylinder and some portion beneath it. As the cylinder is subjected to high working pressure in the subsequent loading phase, the compressive residual stresses counterbalance the tensile stresses generated in the cylinder due to pressurization. This enhances the pressure carrying capacity of the cylinder.

Even though the hydraulic and swage autofrettage are the well-established processes in industries, these processes have certain disadvantages. The hydraulic autofrettage utilizes expensive hydraulic power pack for generating an ultra-high hydraulic pressure inside the cylinder. Also, the process requires skilled operator as the high pressure inside the cylinder needs to be accurately controlled for performing autofrettage. Handling the high pressure may be dangerous and necessitates proper training of the operator. The swage autofrettage requires an expensive hydraulic press and a mandrel made of a material with excessively high strength and hardness. Due to the involvement of explosives, the explosive autofrettage is less practiced in industries. In order to circumvent the difficulties associated with these processes, researchers have proposed alternate methods of achieving autofrettage in the recent years. One of the newly proposed methods is the thermal autofrettage process [10]. The process is simple, inexpensive and easy to handle as compared to the current processes practiced in industries. In thermal autofrettage process, the cylinder is subjected to partial plastic deformation by inducing a radial temperature difference across the wall thickness. The compressive residual thermal stresses are generated in the cylinder upon the removal of the temperature difference. The thermal autofrettage increases the pressure carrying capacity of the cylinder significantly, although it is less as compared to the hydraulic autofrettage. The thermally autofrettaged cylinders can withstand more thermal gradient than the hydraulic autofrettage [11]. Another recently proposed method of autofrettage is rotational autofrettage [25, 26]. In this process, the cylinder is rotated at a sufficiently high angular velocity in order to create a plastically deforming region in the vicinity of the inner wall of the cylinder. When the angular velocity is reduced to zero, the beneficial compressive residual stresses are set up in the cylinder. A combined method of achieving autofrettage by inducing the pressure and temperature difference simultaneously is studied by Shufen and Dixit [21]. The combined process requires relatively less pressure than the hydraulic autofrettage.

In the design of thick-walled cylinders, one of the prime objectives is to increase their fatigue life. This objective can be achieved by autofrettage. The probability of crack initiation at the inner wall of the cylinder is reduced due to compressive residual stresses induced after autofrettage. This slows down the growth of cracks and increases the fatigue life of the cylinder [18, 19, 23]. Sometimes shrink-fitting is combined with autofrettage in order to achieve further enhancement of the fatigue life. The shrink-fit of cylinders alone generates less compressive residual stress compared to autofrettage process at the inside region of a cylinder. The combination of both shrink-fit and autofrettage may provide a more suitable residual stress distribution increasing the performance of the compound cylinder than the autofrettaged monobloc cylinder. Researchers have studied the combination of hydraulic autofrettage and shrink-fit to enhance the residual stress distribution and its impact on the fatigue life. Kapp et al. [13] proposed a multilayer design procedure incorporating shrink-fit of cylindrical layers on a previously autofrettaged monobloc cylinder in order to achieve a very long fatigue life. Parker and Kendall [17] presented a design methodology for combination of hydraulic autofrettage and shrink-fit. Their procedure consists of a sequence involving shrink-fit of two cylinders followed by autofrettage. The procedure reduces the near-bore plastic strain by approximately 50% compared to that occurring in the monobloc autofrettaged cylinder, thereby reducing the impact of Bauschinger effect. The authors achieved an increase of 41% in the fatigue life of the cylinder.

A methodology for design optimization of a three-layered cylindrical vessel for achieving the maximum fatigue life under the combined effects of autofrettage and shrink-fit was presented by Jahed et al. [8] based on actual material behaviour. Authors observed that with the proper combination of operations, a significant enhancement in the fatigue life of the cylinder could be achieved. Some recent studies on the combination of hydraulic autofrettage and shrink-fit of cylinders were carried out by Gexia and Hongzhao [6], Abdelsalam and Sedaghati [1] and Bhatnagar [4]. Gexia and Hongzhao [6] presented analytical solutions for residual stresses in shrink-fit two-layer cylinders after autofrettage based on the actual tensile-compressive stress–strain curve of the material, von Mises yield criterion and plane strain condition. Abdelsalam and Sedaghati [1] presented a design optimization methodology to identify optimal configurations of a two-layer cylinder subjected to different combinations of shrink-fit and autofrettage processes using genetic algorithm and sequential quadratic programming (SQP) optimization techniques. They optimized the thickness of each layer, autofrettage pressure and radial interference for each shrink-fit and autofrettage combination in order to achieve the maximum fatigue life of the compound cylinder. The modelling of autofrettage and shrink-fit compound cylinder with Bauschinger effect was carried out by Bhatnagar [4]. The autofrettage and post-autofrettage machining was validated using FEM on ANSYS. The author concluded that the compounded cylinder provides a high maximum safe pressure, good fatigue life and manufacturing economy in terms of reduced autofrettage pressure leading to the reduction in the maintenance of a hydraulic autofrettage plant.

The enhancement in the pressure carrying capacity of thermally autofrettaged cylinder through shrink-fit was studied by Kamal and Dixit [12]. They considered shrink-fitting of an outer cylindrical layer to a thermally autofrettaged cylinder and analysed the pressure carrying capacity of the compound cylinder. In the present work, the work of Kamal and Dixit [12] is extended to analyse the fatigue life of the thermally autofrettaged cylinder with shrink-fit. Further, the fatigue life of the thermally autofrettaged cylinder with shrink-fit is compared with the fatigue lives of the corresponding thermally autofrettaged as well as non-autofrettaged single/ monobloc cylinders. For the analyses, the stress intensity factor given by Underwood [24] is used. The fatigue life is estimated using Paris law. It is shown that the thermally autofrettaged cylinder with shrink-fit provides larger fatigue life than the corresponding single/monobloc thermally autofrettaged as well as non-autofrettaged cylinder for a safe working pressure. Before carrying out the detailed analysis of fatigue life of the thermally autofrettaged cylinder combined with shrink-fit, the thermal autofrettage process is briefly described in Sect. 1.2.

1.2   The Thermal Autofrettage Process

The thermal autofrettage is a method of producing beneficial compressive residual stresses in cylinders by inducing a radial temperature difference. A sufficiently high temperature difference is created across the wall thickness of the cylinder in order to deform the material at the inside plastically. The outer region of the cylinder gets elastically deformed. On further increasing the temperature difference across the wall thickness of the cylinder, the outer surface is subjected to plastic deformation. This results in an inner plastic zone and an outer plastic zone keeping the intermediate material in the elastically deformed state. On removing the plastically deforming temperature difference across the wall thickness, i.e. after cooling the cylinder to room temperature, beneficial compressive residual stresses are produced in the vicinity of the inner surface of the cylinder. The thermal autofrettage is represented schematically in Fig. 1.1a and the experimental set up for performing thermal autofrettage is shown in Fig. 1.1b [9].

As shown in Fig. 1.1, the outside surface of the cylinder is heated using a ceramic jacketed heater and at the same time, the inner surface is cooled by flowing cold water. This creates a temperature difference across the wall of the cylinder. The temperature difference in the cylinder can be controlled by using a digital temperature controller. For this purpose, the heater is connected in series with a variable autotransformer, an ammeter and a digital temperature controller. The thermocouples from the inner and outer surfaces of the cylinder are connected to the temperature controllers to record the corresponding temperatures. The temperature difference in the cylinder is gradually increased to the desired value to cause the partial plastic deformation within the wall of the cylinder. When the desired value

(a)

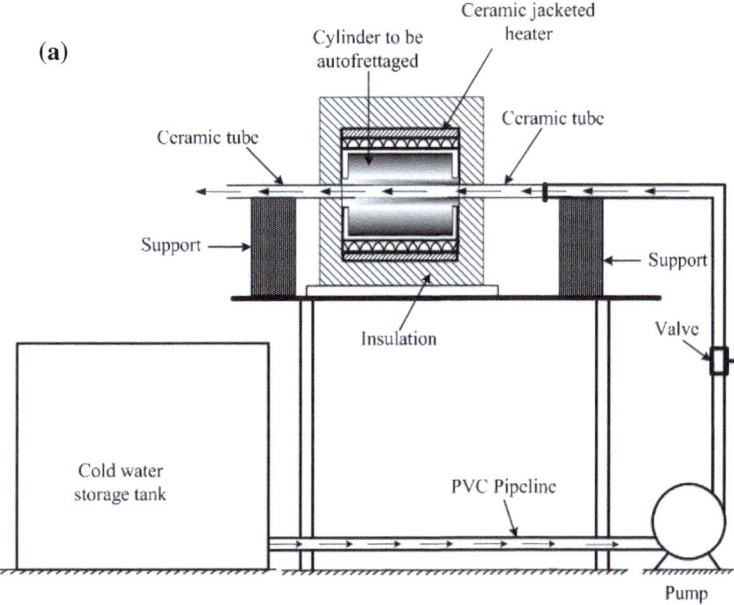

(b)

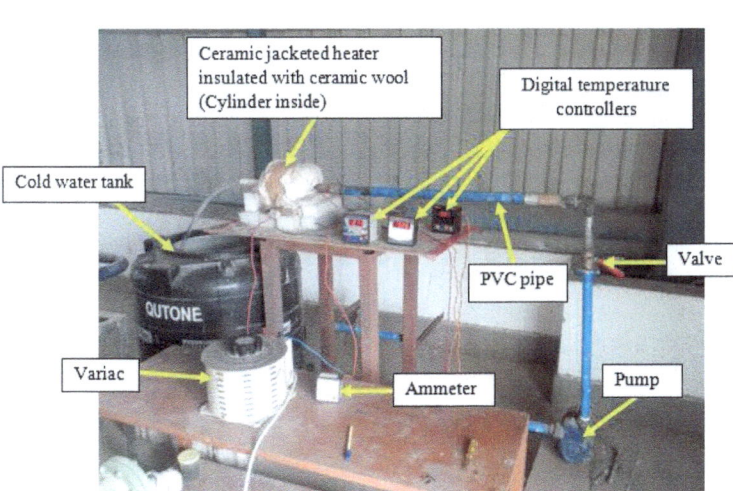

Fig. 1.1 Thermal autofrettage: **a** schematic representation and **b** photograph of experimental set up [9]

of plastically deforming temperature difference is attained, the heater is switched off and cold water is still circulated through the bore till the temperature difference becomes zero. The compressive residual stresses are now set up in the cylinder.

The analytical solutions of the residual stresses induced in the thermally cylinder were obtained by Kamal and Dixit [10]. The analysis is based on the generalized plane strain condition, Tresca yield criterion and its associated flow rule. The use of Tresca yield criterion in the analysis of thermal autofrettage evolves two consecutive plastic zones at the inner portion of the cylinder when there is yielding only at the inner side (first stage of plastic deformation). When there is simultaneous yielding both at the inner and outer wall of the cylinder, then there are two consecutive plastic zones at the inner wall and two consecutive plastic zones at the outer wall (second stage of plastic deformation). The analytical residual stress solutions of Kamal and Dixit [10] are used in the present study and are provided in Appendix A. The present study combines the thermal autofrettage with shrink-fit for the analysis of fatigue life. The stresses due to shrink-fit are provided in Appendix B based on standard textbooks such as Srinath [22].

1.3   Problem Definition

A thick-walled SS304 cylinder with inner radius $a = 10$ mm and outer radius $b = 30$ mm is considered for thermal autofrettage. An outer SS304 cylindrical layer of thickness 10 mm is considered to be shrink-fitted to the thermally autofrettaged cylinder. Hence, the outer radius of the compound cylinder is $z = 40$ mm. The configuration is shown in Fig. 1.2. The material properties of SS304 are as follows [10]: Young's modulus of elasticity, $E = 193$ GPa, yield stress, $\sigma_Y = 205$ MPa, Poisson's ratio, $v = 0.3$, coefficient of thermal expansion, $\alpha = 17.2 \times 10^{-6}/°C$. The thermally autofrettaged SS304 cylinder is subjected to the first stage of plastic deformation with two inner plastic zones propagating outwards and an outer elastic zone. The net residual stress distribution in the compound cylinder is obtained by the superposition of the residual stresses due to thermal autofrettage and shrink-fit. The residual stresses in the outer cylindrical layer are given by stresses due to shrink-fit only. Kamal and Dixit [12] carried out numerical simulations of the thermally autofrettaged cylinder with shrink-fit for different combinations of temperature difference $(T_b - T_a)$ and shrink-fit allowance δ. They obtained the maximum pressure carrying capacity of the thermally autofrettaged cylinder with shrink-fit as 159 MPa for the combination of temperature difference $(T_b - T_a) = 103 \; °C$ and shrink-fit allowance $\delta = 0.027$ mm. Here, the fatigue life analysis of the thermally autofrettaged cylinder combined with shrink-fit is carried out for this combination of the temperature difference and shrink-fit allowance providing the maximum pressure carrying capacity in the cylinder. A comparison of the fatigue life of the compound cylinder with the fatigue lives of the corresponding thermally autofrettaged as well as non-autofrettaged single/monobloc cylinders with inner radius $a = 10$ mm and outer radius $b = 40$ mm is also carried out.

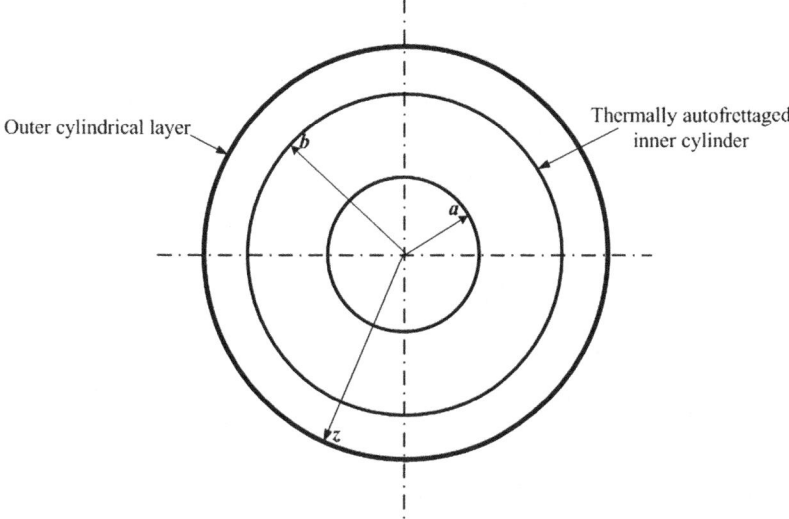

Fig. 1.2 The shrink-fitted compound cylinder

1.4   Fatigue Life Analysis

The fatigue life of thermally autofrettaged cylinder with shrink-fit is analysed. Further, it is compared with the fatigue lives of single/monobloc thermally autofrettaged as well as non-autofrettaged cylinders. For the estimation of fatigue life, the stress intensity factor given by Underwood [24] for straight-fronted, longitudinal crack is employed. The fatigue lives for each case are calculated using Paris law.

1.4.1   Calculation of Stress Intensity Factor

The calculation of stress intensity factors for cracks in thick-walled cylinder is required for the prediction of fatigue life. Before carrying out the detailed calculation of stress intensity factors, the stress intensity factor in general is defined in Sect. 1.4.1.1.

1.4.1.1   Stress Intensity Factor

The stress intensity factor is used to predict the state of stress in the vicinity of the tip of a crack existing in a material due to remote loading or residual stresses. It is usually applied to homogeneous, linear elastic material. It is denoted by the letter K. For the general case, the stress intensity factor K is represented by Sanford [20]

$$K = \beta\sigma\sqrt{\pi l}, \tag{1.1}$$

where β is a parameter that depends on the specimen and crack sizes and geometries and the manner of application of load, σ is the far field stress due to the external loading and l is the crack length. There are three basic modes of fracture—mode-I, mode-II and mode-III. The mode-I fracture corresponds to the one where the crack surfaces are displaced normal to themselves. In mode-II fracture, the crack surfaces are subjected to shear relative to each other in a direction perpendicular to the edge of the crack. Mode-III fracture corresponds to the shearing action parallel to the edge of the crack. To indicate the stress intensity factors corresponding to different modes of fracture, the corresponding subscripts are added to the symbol K. For example, the mode-I stress intensity factor is represented by K_I. In the present study, the mode-I stress intensity factor for a crack existing in a thick-walled cylinder is considered. Mode-I is also called opening mode.

1.4.1.2   The Stress Intensity Factor for Crack in a Thick-Walled Cylinder

The first mode of stress intensity factor for a straight-fronted longitudinal crack of $l/w < 0.25$ (l is the crack depth and w is the wall thickness) under pressure p is given by Underwood [24]

$$K_I = 1.12\sigma_\theta\sqrt{\pi l} + 1.13p\sqrt{\pi l} + 1.12\sigma_{\theta R}\sqrt{\pi l}, \tag{1.2}$$

where σ_θ is the Lamé hoop stress and $\sigma_{\theta R}$ is the residual hoop stress. The crack geometry is shown in Fig. 1.3. Under a cyclic pressure, the range of stress intensity, ΔK_I, is

$$\Delta K_I = K_{Imax} - K_{Imin}, \tag{1.3}$$

where K_{Imax} and K_{Imin} are given by

$$K_{Imax} = 1.12\sigma_{\theta max}\sqrt{\pi l} + 1.13p_{max}\sqrt{\pi l} + 1.12\sigma_{\theta R}\sqrt{\pi l}, \tag{1.4}$$

$$K_{Imin} = 1.12\sigma_{\theta min}\sqrt{\pi l} + 1.13p_{min}\sqrt{\pi l} + 1.12\sigma_{\theta R}\sqrt{\pi l}. \tag{1.5}$$

In Eqs. (1.4) and (1.5), p_{max} and p_{min} are the maximum and minimum pressure in the cycle. For large compressive residual stresses, K_{Imin} may result in a negative value. In that case, K_{Imin} is set to zero providing $\Delta K_I = K_{Imax}$. Then,

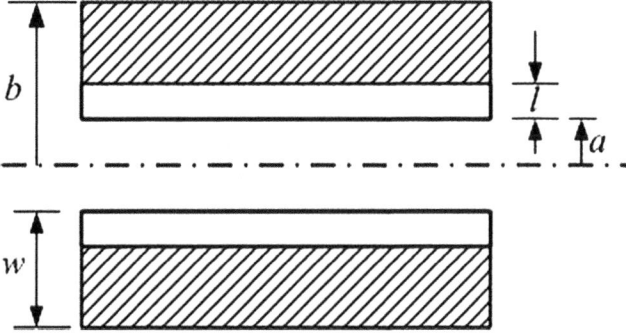

Fig. 1.3 A straight-fronted longitudinal crack emanating from the inner wall of a thick cylinder showing crack geometry parameters [9]

$$\Delta K_{\mathrm{I}} = 1.12\Delta\sigma_\theta \sqrt{\pi l} + 1.13\Delta p \sqrt{\pi l} + 1.12\sigma_{\theta R} \sqrt{\pi l}. \tag{1.6}$$

The range of hoop stress at any radius, r, is given by the Lamé hoop stress distribution as

$$\Delta\sigma_\theta = \frac{\Delta p\left(1 + \frac{b^2}{r^2}\right)}{(W^2 - 1)}, \tag{1.7}$$

where b is the outer radius of the cylinder and W is the wall thickness ratio given by b/a, where a is the inner radius of the cylinder. Inserting Eq. (1.7) into Eq. (1.6), the stress intensity factor in non-dimensional form is given by

$$\frac{\Delta K_{\mathrm{I}}}{\Delta K_o} = 1.12\left\{\frac{\left(1 + \frac{b^2}{r^2}\right)}{(W^2 - 1)} + 1.009 + \frac{\sigma_{\theta R}}{\Delta p}\right\}, \tag{1.8}$$

where $\Delta K_o = \Delta p \sqrt{\pi l}$. For non-autofrettaged cylinder, $\sigma_{\theta R}$ is zero in Eq. (1.8).

The stress intensity factor is calculated as a function of crack depth ratio (l/w). The crack depth ratio is defined as

$$\frac{l}{w} = \frac{r - a}{b - a}. \tag{1.9}$$

Corresponding to a particular value of l/w, the radial position, r can be obtained from Eq. (1.9), which is substituted in Eq. (1.8). In the following subsections, the stress intensity factor is calculated for different cases.

Thermally Autofrettaged Cylinder with Shrink-Fit

The non-dimensional stress intensity factor for the thermally autofrettaged cylinder with shrink-fit defined in Sect. 4.1.2 is calculated for four different pressures −96 MPa (the yield onset pressure of the non-autofrettaged monobloc cylinder), 110 MPa (the intermediate pressure between the yield onset pressure of non-autofrettaged monobloc cylinder and the thermally autofrettaged monobloc cylinder), 134 MPa (the yield onset pressure of the thermally autofrettaged monobloc cylinder) and 159 MPa (the yield onset pressure of the thermally autofrettaged cylinder with shrink-fit). It is considered that the minimum pressure in the cylinder during operation is zero. A straight-fronted longitudinal crack is considered at the inner wall of the compound cylinder as shown in Fig. 1.3. The thickness of inner cylinder is 20 mm. The allowable final crack depth is taken as per ASME boiler and pressure vessel code [2]. The ASME boiler and pressure vessel code [2], Section VIII—Division 3 regulates the rules for the construction of high-pressure vessels. The Article KD-4 of ASME boiler and pressure vessel code, Section VIII—Division 3 presents a fracture mechanics design approach of pressure vessels. It utilizes the principles of linear fracture mechanics to develop the criterion for calculating the number of design cycles to propagate the cracks to the critical crack depth and the maximum allowable depth. As per the ASME code standard, the allowable final crack depth shall not exceed 25% of the inner layer thickness. Thus, in the present case, the allowable final crack depth is taken as 5 mm. The wall thickness ratio of the compound cylinder to be used in Eq. (1.8) is defined as, $W = z/a = 4$ with reference to Fig. 1.2. The stress intensity factor for the compound cylinder is calculated using Eq. (1.8) for the different pressures and is shown in Fig. 1.4 as a function of (l/w). It is observed that the stress intensity factor increases up to the radius of elastic–plastic interface and thereafter it starts decreasing gradually. Also, the stress intensity factor increases with the increase in working pressure following the same trend.

Thermally Autofrettaged Monobloc Cylinder

A monobloc SS304 cylinder is considered for thermal autofrettage inducing a temperature difference of 122 °C across the wall thickness for achieving the maximum pressure carrying capacity (134 MPa). The residual stress distribution in the thermally autofrettaged cylinder can be evaluated from the expressions presented in Appendix A. The crack geometry parameters for this cylinder correspond to Fig. 1.3. The non-dimensional stress intensity factor for this cylinder is calculated using Eq. (1.8) for three different pressures—96, 110 and 134 MPa and is shown in Fig. 1.5 as a function of (l/w). The stress intensity factor reaches the maximum value at the interface radius between the plastic zones I and II and thereafter it starts decreasing gradually.

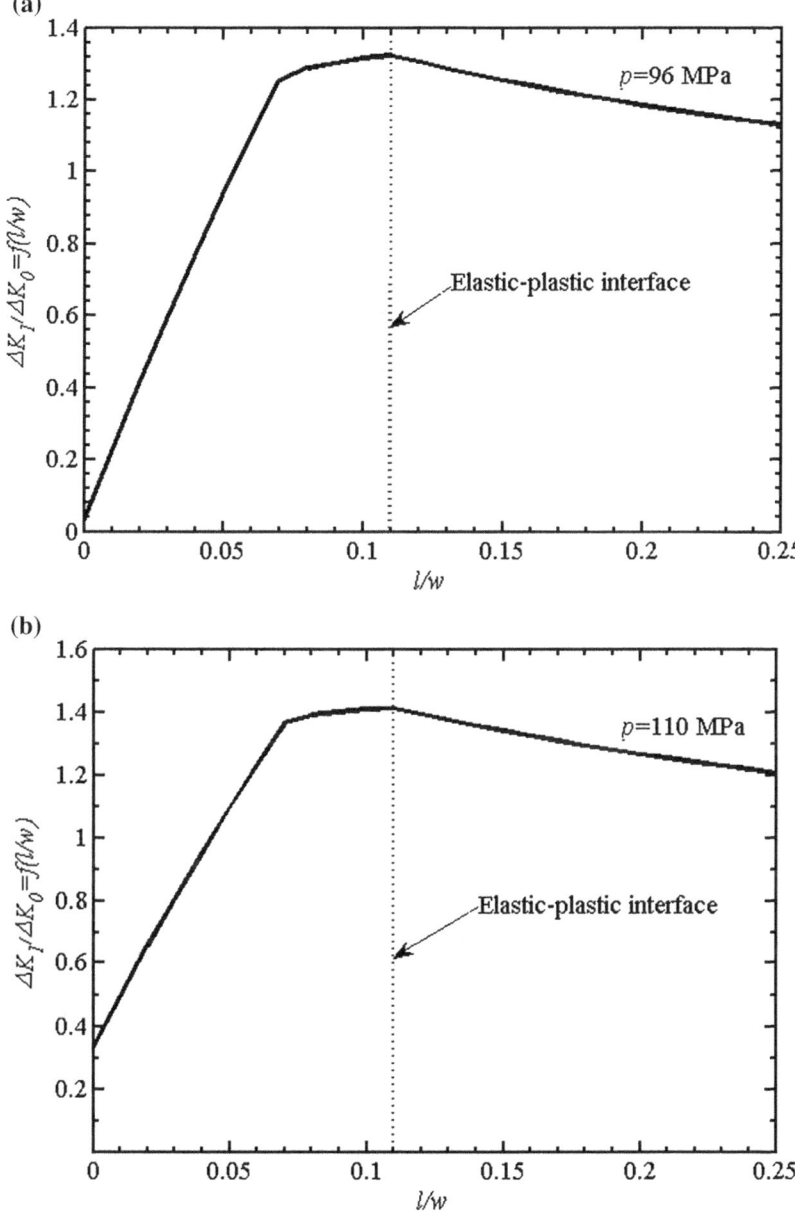

Fig. 1.4 Stress intensity factors for a straight-fronted longitudinal crack in the autofrettaged cylinder with shrink-fit for different working pressures: **a** $p = 96$ MPa, **b** $p = 110$ MPa, **c** $p = 134$ MPa, **d** $p = 159$ MPa

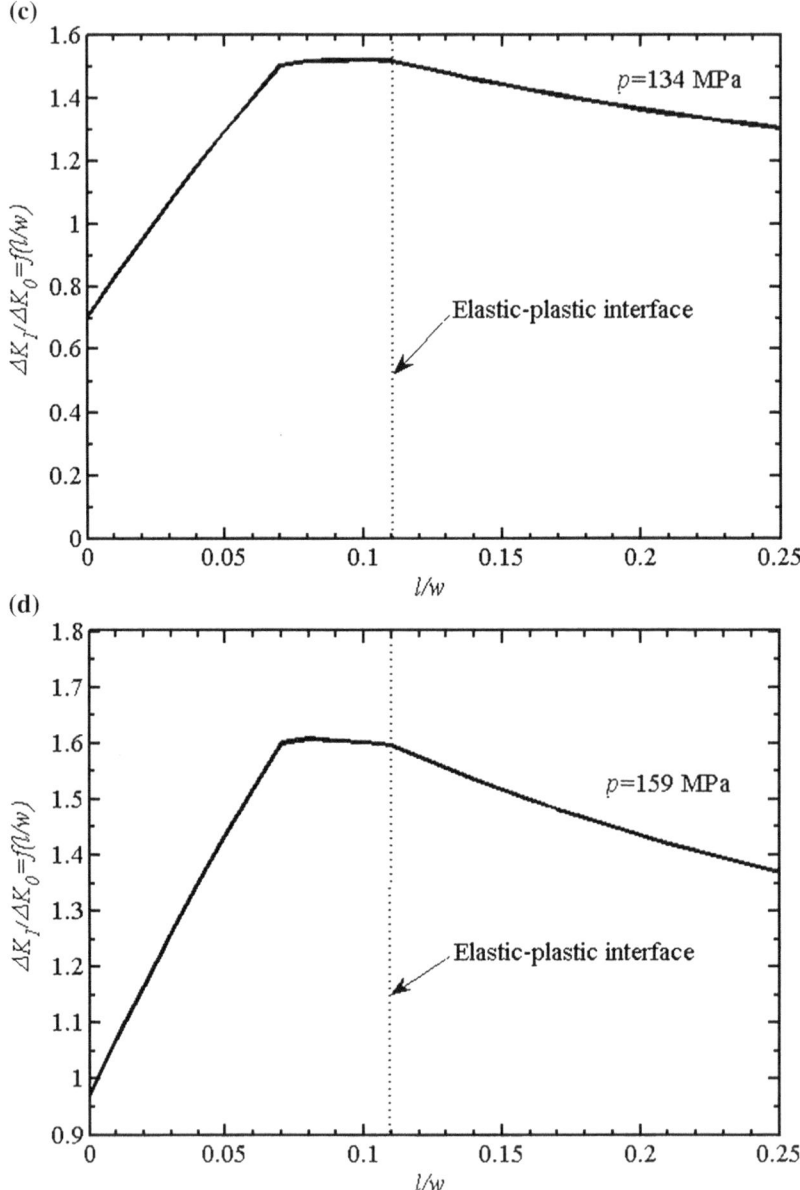

Fig. 1.4 (continued)

(a)

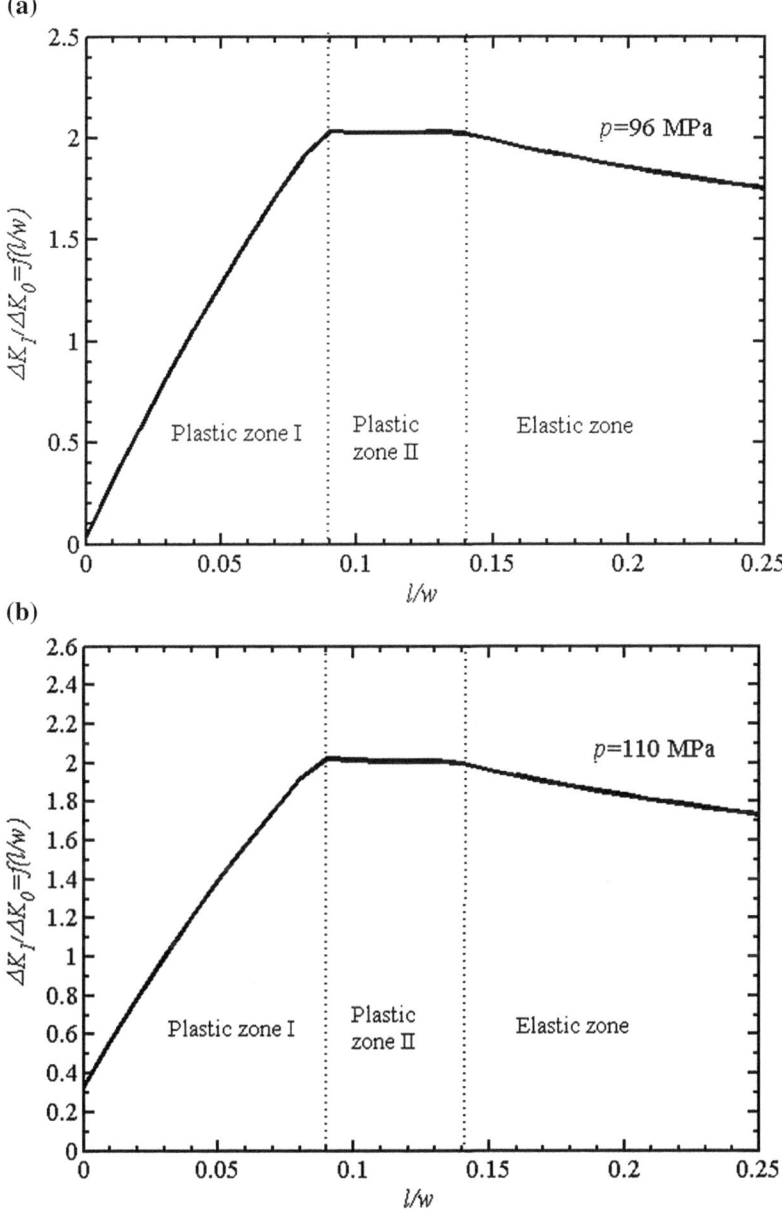

(b)

Fig. 1.5 Stress intensity factors for a straight-fronted longitudinal crack in the autofrettaged SS304 monobloc cylinder for different working pressures: **a** $p = 96$ MPa, **b** $p = 110$ MPa, **c** $p = 134$ MPa

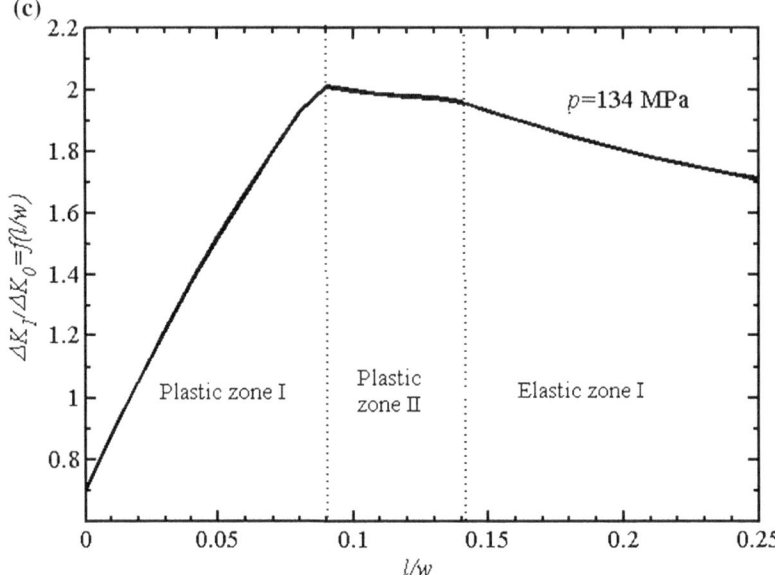

Fig. 1.5 (continued)

Non-autofrettaged Monobloc Cylinder

The non-autofrettaged cylinder can withstand the maximum pressure of 96 MPa. The non-dimensional stress intensity factors for this cylinder at the maximum pressure carrying capacity are calculated from Eq. (1.8) putting $\sigma_{\theta R} = 0$. The distribution is shown in Fig. 1.6 as a function of (l/w) for the crack geometry shown in Fig. 1.3. It is observed that the stress intensity factor becomes the maximum at the inner radius and then decreases gradually. The high stress intensity factor in the cylinder causes the crack propagation faster reducing its fatigue life. For increasing the fatigue life, it is desirable that the stress intensity factor along the crack length should be as low as possible. From Figs. 1.4 and 1.6, it is observed that the maximum stress intensity factor in the thermally autofrettaged cylinder for the pressures of 96, 110, 134 and 159 MPa is reduced by 84.6, 71.42, 57.89 and 50%, respectively, as compared to the maximum stress intensity factor in the non-autofrettaged monobloc cylinder. Similarly comparing Figs. 1.5 and 1.6, it is observed that the stress intensity factor in the thermally autofrettaged monobloc cylinder for the pressures of 96, 110 and 134 MPa is reduced by about 20% as compared to the non-autofrettaged monobloc cylinder. This increases the fatigue life of the thermally autofrettaged cylinder with shrink-fit and thermally autofrettaged monobloc cylinder with respect to the corresponding non-autofrettaged monobloc cylinder for different safe working pressures.

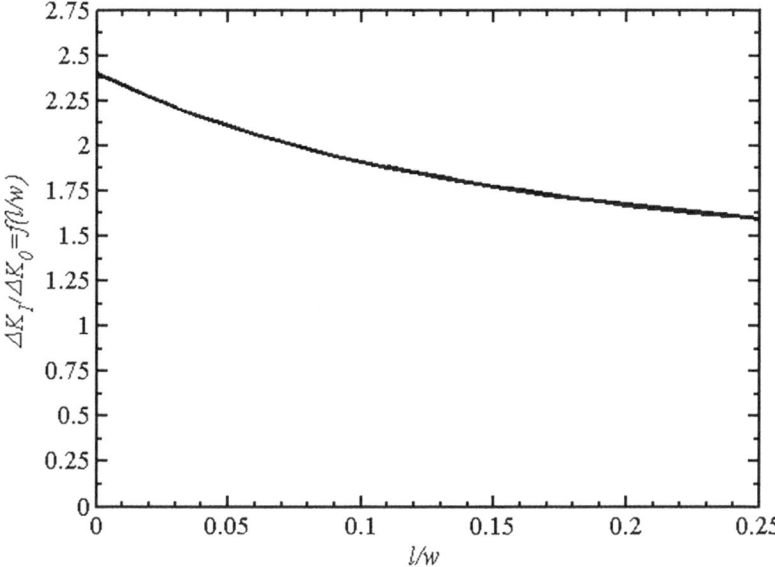

Fig. 1.6 Stress intensity factors for a straight-fronted longitudinal crack in the non-autofrettaged monobloc SS304 cylinder for 96 MPa

1.4.2 Calculation of Fatigue Life

Paris et al. [16] suggested using the range of stress intensity factor ΔK to characterize the rate of crack propagation per cycle dl/dN under cyclic loading. By plotting the rate of crack propagation per cycle against the range of stress intensity factor for a number of alloys, they found that the plot provides straight lines on log-log scales. Thus, they obtained the following relation, which is known as Paris law:

$$\frac{dl}{dN} = A(\Delta K_{\mathrm{I}})^{m}, \qquad (1.10)$$

where A and m are material constants. Equation (1.10) makes it possible to quantify the prediction of fatigue life for a crack of a certain size. As $\Delta K_{\mathrm{I}}/\Delta K_o = f(l/w)$,

$$\Delta K_{\mathrm{I}} = \Delta K_o f\left(\frac{l}{w}\right) = \Delta p \sqrt{\pi l} f\left(\frac{l}{w}\right). \qquad (1.11)$$

Using Eq. (1.11) in Eq. (1.10),

$$\frac{dl}{dN} = A(\pi)^{\frac{m}{2}}(\Delta p)^{m}(l)^{\frac{m}{2}}\{f(l/w)\}^{m}. \qquad (1.12)$$

Integrating Eq. (1.12), the fatigue life is obtained as

$$N_f = \frac{w^{1-\frac{m}{2}}}{A(\pi)^{\frac{m}{2}}(\Delta p)^m} \int_{(l/w)_i}^{(l/w)_f} \frac{\mathrm{d}(l/w)}{(l/w)^{\frac{m}{2}}[f(l/w)]^m}. \tag{1.13}$$

The lower limit of the integration $(l/w)_i$ is the initial crack depth ratio and the upper limit of the integration $(l/w)_f$ is the final crack depth ratio in Eq. (1.13). The integral can be evaluated numerically using Simpson's 1/3rd rule. Simpson's rule approximates the integral of a function using quadratic polynomials. The interval of the integral is divided into an even number of subintervals in Simpson's rule. If one needs to approximate an integral in the interval from a to b with n number of subintervals, then the step size is given by $(b - a)/n$.

The fatigue life of the cylinder for different cases is calculated using Eq. (1.13). The initial crack depth ratio is taken as 0.001 [14] and the final crack depth ratio is taken as 0.25 as per the ASME boiler and pressure vessel code [2]. For SS304 (austenitic steel), the material constants in Eq. (1.12) are taken as $A = 5.5 \times 10^{-12}$ and $m = 3.25$ [3]. In evaluating the integrand of Eq. (1.12), the values of $f(l/w)$ are calculated from Figs. 1.3, 1.4 and 1.5 for the respective cases fitting polynomial curves. The numerical integration is carried out in a step size of 0.00498. The summary of the results is presented in Table 1.1.

It is observed that the fatigue life of the thermally autofrettaged cylinder is significantly increased by shrink-fitting. The fatigue of the cylinders is dependent on the working pressures. With the increase of working pressure, the fatigue life is reduced. For the working pressure of 96 MPa, the fatigue life of the thermally autofrettaged cylinder with shrink-fit increases to 2,413,854,831 cycles, which is 176,464 times higher than that of the non-autofrettaged monobloc cylinder. With respect to the thermally autofrettaged monobloc cylinder, the increase in fatigue life is 1,769,593 cycles. For 110 MPa, the enhancement in fatigue life is 57,422 cycles with respect to the thermally autofrettaged monobloc cylinder. The fatigue life of the thermally autofrettaged cylinder with shrink-fit is increased by 3,572 cycles than the thermally autofrettaged monobloc cylinder when the cylinder is under the working pressure of 134 MPa. Thus, the thermally autofrettaged cylinder with shrink-fit has the maximum fatigue life for a fixed safe working pressure. At its

Table 1.1 Fatigue lives of cylinders for different ranges of working pressures

Working pressure (MPa)	Fatigue lives (cycles)		
	Thermally autofrettaged cylinder with shrink-fit	Thermally autofrettaged monobloc cylinder	Non-autofrettaged monobloc cylinder
96	2,413,854,831	2,412,085,238	13,679
110	3,363,670	3,306,248	–
134	188,934	185,362	–
159	39,674	–	–

maximum pressure carrying capacity (159 MPa), the fatigue life of this cylinder is 39,674 cycles which is 2.9 times the fatigue life of the non-autofrettaged monobloc cylinder operating at its maximum working pressure (96 MPa).

1.5   Conclusions

In this work, a study on enhancing the fatigue life of the thermally autofrettaged SS304 cylinder with shrink-fit is carried out. The thermal autofrettage is a recently proposed autofrettage method [10]. Many researchers have studied the fatigue life of hydraulically autofrettaged thick-walled cylinder combined with shrink-fit. In this work, the fatigue life of thick-walled cylinder combining thermal autofrettage with shrink-fit is theoretically studied for the first time. For analysing the fatigue lives, the first mode of stress intensity factor given by Underwood [24] is used for a straight-fronted longitudinal crack. The fatigue lives are calculated for different safe working pressures using Paris law. Further, the fatigue lives of the corresponding single/monobloc thermally autofrettaged as well as non-autofrettaged cylinders are evaluated for the same safe working pressures and compared with the corresponding fatigue lives of the thermally autofrettaged cylinder with shrink-fit.

The following conclusions are drawn from the present study:

- The thermally autofrettaged cylinder with shrink-fit provides a very high enhancement in the fatigue life for different working pressures as compared to the fatigue life of the non-autofrettaged monobloc cylinder at its maximum pressure carrying capacity.
- Corresponding to the yield onset pressure of the non-autofrettaged monobloc SS304 cylinder (96 MPa), the thermally autofrettaged SS304 cylinder with shrink-fit enhances the fatigue life by 176,464 times with respect to the non-autofrettaged monobloc cylinder.
- Corresponding to the yield onset pressure of the thermally autofrettaged monobloc SS304 cylinder (134 MPa), the thermally autofrettaged SS304 cylinder provides a fatigue life of 188,934 cycles. A non-autofrettaged cylinder is not able to sustain this pressure.
- The thermally autofrettaged SS304 cylinder with shrink-fit has the maximum pressure carrying capacity of 159 MPa.
- The thermally autofrettaged cylinder with shrink-fit provides 1,769,593 cycles higher fatigue life than the corresponding thermally autofrettaged monobloc cylinder for the working pressure of 96 MPa. This increase is insignificant. Hence, at low pressure, shrink-fit is not advantageous.
- For the working pressure of 134 MPa, the thermally autofrettaged cylinder with shrink-fit has 3,572 cycles higher fatigue life than the corresponding thermally autofrettaged monobloc cylinder. This is about 2% increase in the fatigue life.
- The thermally autofrettaged cylinder with shrink-fit is able to sustain 159 MPa pressure with a fatigue life of 39,674 cycles. Without shrink-fit, this much high pressure cannot be sustained.

Overall, the thermally autofrettaged cylinder with shrink-fit provides improved fatigue life for a particular safe working pressure. The fatigue life of the thermally autofrettaged cylinder with shrink-fit decreases as the working pressure increases. In future, fatigue test experiments may be carried out in order to validate these estimations. Different metallic materials may be considered for both theoretical and experimental studies to get a better insight of the influence of autofrettage on fatigue life.

Appendix 1: Residual Stress Distributions During Thermal Autofrettage of a Thick-Walled Cylinder

A thick-walled cylinder with inner radius a and outer radius b is considered. The cylinder is subjected to a temperature difference $(T_b - T_a)$ for achieving thermal autofrettage. As $(T_b - T_a)$ crosses the temperature difference required for initial yielding, the cylinder undergoes the first stage of elastic–plastic deformation causing the material at the inner wall to deform plastically. If $(T_b - T_a)$ is increased further, at certain temperature difference, the second stage of elastic–plastic deformation occurs in the cylinder causing the simultaneous plastic deformation both at the inner and outer wall. The two stages of deformations in the cylinder with different elastic and plastic zones developed as per Tresca yield criterion are shown in Fig. 1.1. After the release of temperature difference, $(T_b - T_a)$, residual stresses are generated in the cylinder. Referring to Fig. 1.1a, the residual stresses generated in different zones during the first stage of elastic–plastic deformation under the condition of generalized plane strain are given as follows [10] (Fig. 1.7):

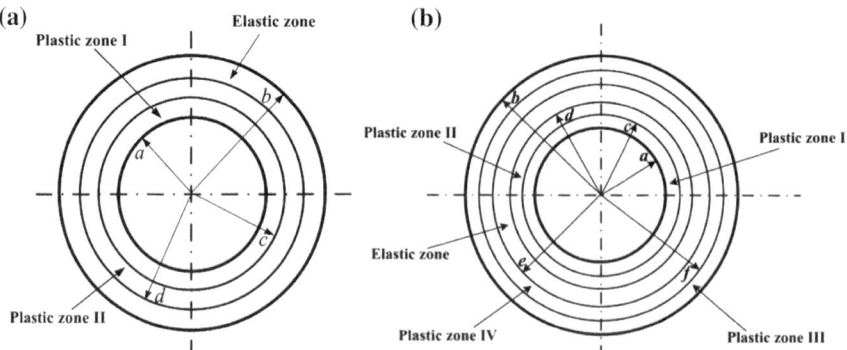

Fig. 1.7 Different elastic and plastic zones in the cylinder during **a** the first stage and **b** second stage of elastic–plastic deformation in thermal autofrettage. With permission from Kamal and Dixit [10]. Copyright 2015, ASME

Plastic zone I, $a \leq r \leq c$:

$$\sigma_r^{\mathrm{I}} = k_1 \sigma_Y \ln r + C_3 + \frac{E\alpha(T_b - T_a)}{2(1-v)\ln\left(\frac{b}{a}\right)} \left\{ \ln\left(\frac{r}{a}\right) - \ln\left(\frac{b}{a}\right)\left(1 - \frac{a^2}{r^2}\right)\frac{b^2}{b^2 - a^2} \right\},$$

(1.14)

$$\sigma_\theta^{\mathrm{I}} = k_1 \sigma_Y (1 + \ln r) + C_3 + \frac{E\alpha(T_b - T_a)}{2(1-v)\ln\left(\frac{b}{a}\right)} \left\{ 1 + \ln\left(\frac{r}{a}\right) - \ln\left(\frac{b}{a}\right)\left(1 + \frac{a^2}{r^2}\right)\frac{b^2}{b^2 - a^2} \right\},$$

(1.15)

$$\sigma_z^{\mathrm{I}} = k_1 \sigma_Y (1 + \ln r) + C_3 + \frac{E\alpha(T_b - T_a)}{2(1-v)\ln\left(\frac{b}{a}\right)} \left\{ 1 + 2\ln\left(\frac{r}{a}\right) - \ln\left(\frac{b}{a}\right)\frac{2b^2}{b^2 - a^2} \right\}.$$

(1.16)

Plastic zone II, $c \leq r \leq d$:

$$\sigma_r^{\mathrm{II}} = C_5 r^{-1 + \sqrt{2(1-v)}} + C_6 r^{-1 - \sqrt{2(1-v)}} + \frac{k_1 \sigma_Y}{(2v-1)} + \frac{E\alpha T_a}{(2v-1)}$$
$$+ \frac{E\alpha(T_b - T_a)}{(2v-1)\ln\left(\frac{b}{a}\right)} \left\{ \ln\left(\frac{r}{a}\right) - \frac{2v+1}{2v-1} \right\}$$
$$- \frac{E\varepsilon_0}{(2v-1)} + \frac{E\alpha(T_b - T_a)}{2(1-v)\ln\left(\frac{b}{a}\right)}$$
$$\left\{ \ln\left(\frac{r}{a}\right) - \ln\left(\frac{b}{a}\right)\left(1 - \frac{a^2}{r^2}\right)\frac{b^2}{b^2 - a^2} \right\},$$

(1.17)

$$\sigma_\theta^{\mathrm{II}} = C_5 \sqrt{2(1-v)}\, r^{\sqrt{2(1-v)}-1} - C_6 \sqrt{2(1-v)}\, r^{-\sqrt{2(1-v)}-1} + \frac{k_1 \sigma_Y}{(2v-1)} + \frac{E\alpha T_a}{(2v-1)}$$
$$+ \frac{E\alpha(T_b - T_a)}{(2v-1)\ln\left(\frac{b}{a}\right)} \left\{ \ln\left(\frac{r}{a}\right) - \frac{2}{2v-1} \right\} - \frac{E\varepsilon_0}{(2v-1)}$$
$$+ \frac{E\alpha(T_b - T_a)}{2(1-v)\ln\left(\frac{b}{a}\right)} \left\{ 1 + \ln\left(\frac{r}{a}\right) - \ln\left(\frac{b}{a}\right)\left(1 + \frac{a^2}{r^2}\right)\frac{b^2}{b^2 - a^2} \right\},$$

(1.18)

$$\sigma_z^{\mathrm{II}} = C_5 r^{-1 + \sqrt{2(1-v)}} + C_6 r^{-1 - \sqrt{2(1-v)}} + \left(\frac{2v}{2v-1}\right) k_1 \sigma_Y + \frac{E\alpha T_a}{(2v-1)} + \frac{E\alpha(T_b - T_a)}{(2v-1)\ln\left(\frac{b}{a}\right)}$$
$$\left\{ \ln\left(\frac{r}{a}\right) - \frac{2v+1}{2v-1} \right\} - \frac{E\varepsilon_0}{(2v-1)} + \frac{E\alpha(T_b - T_a)}{2(1-v)\ln\left(\frac{b}{a}\right)}$$
$$\left\{ 1 + 2\ln\left(\frac{r}{a}\right) - \ln\left(\frac{b}{a}\right)\frac{2b^2}{b^2 - a^2} \right\}.$$

(1.19)

Elastic zone, $d \leq r \leq b$:

$$
\sigma_r^{el} = \frac{E\alpha}{2(1-v)} \frac{(T_b - T_a)}{\ln\left(\frac{b}{a}\right)} \left[\ln\left(\frac{b}{a}\right) - \ln\left(\frac{b}{a}\right)\left(1 - \frac{a^2}{r^2}\right) \frac{b^2}{b^2 - a^2} \right.
$$
$$
\left. - \left\{ \frac{d^2}{b^2 + d^2(2v-1)} \right\} \left\{ \ln\left(\frac{b}{a}\right)(2v-1) - v - \ln\left(\frac{d}{a}\right) \right\} \left(1 - \frac{b^2}{r^2}\right) \right]
$$
$$
+ \left\{ \frac{d^2}{b^2 + d^2(2v-1)} \right\} k_1 \sigma_Y \left(1 - \frac{b^2}{r^2}\right) + \left\{ \frac{d^2}{b^2 + d^2(2v-1)} \right\} E\alpha T_a
$$
$$
- \left\{ \frac{d^2}{b^2 + d^2(2v-1)} \right\} E\varepsilon_0 \left(1 - \frac{b^2}{r^2}\right),
$$
(1.20)

$$
\sigma_\theta^{el} = \frac{E\alpha}{2(1-v)} \frac{(T_b - T_a)}{\ln\left(\frac{b}{a}\right)} \left[\ln\left(\frac{b}{a}\right) - \ln\left(\frac{b}{a}\right)\left(1 + \frac{a^2}{r^2}\right) \frac{b^2}{b^2 - a^2} \right.
$$
$$
\left. - \left\{ \frac{d^2}{b^2 + d^2(2v-1)} \right\} \left\{ \ln\left(\frac{b}{a}\right)(2v-1) - v - \ln\left(\frac{d}{a}\right) \right\} \left(1 + \frac{b^2}{r^2}\right) \right]
$$
$$
+ \left\{ \frac{d^2}{b^2 + d^2(2v-1)} \right\} k_1 \sigma_Y \left(1 + \frac{b^2}{r^2}\right) + \left\{ \frac{d^2}{b^2 + d^2(2v-1)} \right\} E\alpha T_a \left(1 + \frac{b^2}{r^2}\right)
$$
$$
- \left\{ \frac{d^2}{b^2 + d^2(2v-1)} \right\} E\varepsilon_0 \left(1 + \frac{b^2}{r^2}\right),
$$
(1.21)

$$
\sigma_z^{el} = \frac{E\alpha}{2(1-v)} \frac{(T_b - T_a)}{\ln\left(\frac{b}{a}\right)} \left[1 + 2\ln\left(\frac{r}{a}\right) - \ln\left(\frac{b}{a}\right) \frac{2b^2}{b^2 - a^2} + 2v\ln\left(\frac{b}{r}\right) \right.
$$
$$
\left. -v - \left\{ \frac{2vd^2}{b^2 + d^2(2v-1)} \right\} \left\{ \ln\left(\frac{b}{a}\right)(2v-1) - v - \ln\left(\frac{d}{a}\right) \right\} \right]
$$
$$
+ \left\{ \frac{2vd^2}{b^2 + d^2(2v-1)} \right\} k_1 \sigma_Y + \left\{ \frac{d^2 - b^2}{b^2 + d^2(2v-1)} \right\} E\alpha T_a
$$
$$
- \left\{ \frac{d^2 - b^2}{b^2 + d^2(2v-1)} \right\} E\varepsilon_0 - E\alpha(T_b - T_a) \frac{\ln\left(\frac{r}{a}\right)}{\ln\left(\frac{b}{a}\right)}.
$$
(1.22)

In the above equations, a is the inner radius, b is the outer radius, c is the interface radius between the plastic zone I and II and d is the interface radius between the plastic zone II and the elastic zone. The radii, c and d are obtained by using the boundary conditions of vanishing radial stress at the inner radius and $\sigma_\theta^{(\text{plastic zone II})} = \sigma_z^{(\text{plastic zone II})}$ at $r = c$ as described in Kamal and Dixit [10].

The constants C_3, C_5 and C_6 are given by

$$C_5 = Q + \frac{2vb^2}{(2v-1)\{b^2+d^2(2v-1)\}} d^{1-\sqrt{2(1-v)}} \left\{ \frac{1-v+v\sqrt{2(1-v)}}{2v\sqrt{2(1-v)}} \right\} E\varepsilon_0,$$

(1.23)

$$C_6 = P - \frac{2vb^2}{(2v-1)\{b^2+d^2(2v-1)\}} \left\{ \frac{1-v-v\sqrt{2(1-v)}}{d^{-1-\sqrt{2(1-v)}}2v\sqrt{2(1-v)}} \right\} E\varepsilon_0,$$ (1.24)

where

$$P = \frac{E\alpha(T_b - T_a)}{(2v-1)\ln\left(\frac{b}{a}\right)} \left\{ \frac{1-v-v\sqrt{2(1-v)}}{d^{-1-\sqrt{2(1-v)}}2v\sqrt{2(1-v)}} \right\}$$

$$\left\{ \ln\left(\frac{d}{a}\right) - \frac{2v+1}{2v-1} - \frac{(1+v)}{1-v-v\sqrt{2(1-v)}} \right\}$$

$$- \frac{E\alpha}{2(1-v)} \frac{(T_b-T_a)}{\ln\left(\frac{b}{a}\right)} \left\{ \frac{1-v-v\sqrt{2(1-v)}}{d^{-1-\sqrt{2(1-v)}}2v\sqrt{2(1-v)}} \right\}$$

$$\left[\ln\left(\frac{b}{d}\right) + \left\{ \frac{b^2-d^2}{b^2+d^2(2v-1)} \right\} \left\{ \ln\left(\frac{b}{a}\right)(2v-1) - v - \ln\left(\frac{d}{a}\right) \right\} \right]$$ (1.25)

$$+ \frac{2vb^2}{(2v-1)\{b^2+d^2(2v-1)\}} \left\{ \frac{1-v-v\sqrt{2(1-v)}}{d^{-1-\sqrt{2(1-v)}}2v\sqrt{2(1-v)}} \right\} k_1\sigma_Y$$

$$+ \frac{2vb^2}{(2v-1)\{b^2+d^2(2v-1)\}} \left\{ \frac{1-v-v\sqrt{2(1-v)}}{d^{-1-\sqrt{2(1-v)}}2v\sqrt{2(1-v)}} \right\} E\alpha T_a,$$

and

$$Q = -Pd^{-2\sqrt{2(1-v)}} \left\{ \frac{1-v+v\sqrt{2(1-v)}}{1-v-v\sqrt{2(1-v)}} \right\} - \frac{E\alpha(T_b-T_a)}{(2v-1)\ln\left(\frac{b}{a}\right)}$$

$$\frac{(1+v)}{d^{-1+\sqrt{2(1-v)}}\{1-v-v\sqrt{2(1-v)}\}}.$$ (1.26)

$$C_3 = R + \left[\frac{\frac{2vb^2c^{-1+\sqrt{2(1-v)}}}{(2v-1)\{b^2+d^2(2v-1)\}} d^{1-\sqrt{2(1-v)}} \left\{ \frac{1-v+v\sqrt{2(1-v)}}{2v\sqrt{2(1-v)}} \right\}}{- \frac{2vb^2c^{-1-\sqrt{2(1-v)}}}{(2v-1)\{b^2+d^2(2v-1)\}} \left\{ \frac{1-v-v\sqrt{2(1-v)}}{d^{-1-\sqrt{2(1-v)}}2v\sqrt{2(1-v)}} \right\} - \frac{1}{(2v-1)}} \right] E\varepsilon_0,$$

(1.27)

where

$$R = Qc^{-1+\sqrt{2(1-v)}} - k_1\sigma_Y \ln c + Pc^{-1-\sqrt{2(1-v)}} + \frac{k_1\sigma_Y}{(2v-1)} + \frac{E\alpha T_a}{(2v-1)}$$
$$+ \frac{E\alpha(T_b - T_a)}{(2v-1)\ln\left(\frac{b}{a}\right)}\ln\left(\frac{c}{a}\right) - \frac{2E\alpha(T_b - T_a)}{(2v-1)^2\ln\left(\frac{b}{a}\right)} - \frac{E\alpha(T_b - T_a)}{(2v-1)\ln\left(\frac{b}{a}\right)}, \tag{1.28}$$

The constant axial strain ε_0 is given by

$$\varepsilon_0 = \frac{k_1\sigma_Y}{AE}\left\{\ln c\frac{c^2}{2} - \ln a\frac{a^2}{2} + \frac{1}{4}(c^2 - a^2) + \frac{v(d^2 - c^2)}{2v-1} + \frac{vd^2(b^2 - d^2)}{b^2 + d^2(2v-1)}\right\}$$
$$+ \frac{R}{AE}\left(\frac{c^2 - a^2}{2}\right) + \frac{Q}{AE}\left\{\frac{d^{1+\sqrt{2(1-v)}} - c^{1+\sqrt{2(1-v)}}}{1+\sqrt{2(1-v)}}\right\}$$
$$+ \frac{P}{AE}\left\{\frac{d^{1-\sqrt{2(1-v)}} - c^{1-\sqrt{2(1-v)}}}{1-\sqrt{2(1-v)}}\right\} + \frac{\alpha T_a}{A}\left[\frac{d^2 - c^2}{2(2v-1)} - \frac{(b^2 - d^2)^2}{2\{b^2 + d^2(2v-1)\}}\right]$$
$$+ \frac{\alpha(T_b - T_a)}{(2v-1)\ln\left(\frac{b}{a}\right)A}\left\{\ln\left(\frac{d}{a}\right)\frac{d^2}{2} - \ln\left(\frac{c}{a}\right)\frac{c^2}{2} - \frac{1}{2}(d^2 - c^2)\frac{6v+1}{2(2v-1)}\right\}$$
$$+ \frac{\alpha}{2(1-v)}\frac{(T_b - T_a)}{\ln\left(\frac{b}{a}\right)A}\left[\begin{array}{l} -vd^2\ln\left(\frac{b}{d}\right) - \left\{\frac{vd^2(b^2 - d^2)}{b^2 + d^2(2v-1)}\right\}\left\{\ln\left(\frac{b}{a}\right)(2v-1) - v - \ln\left(\frac{d}{a}\right)\right\} \\ -2(1-v)\left\{\ln\left(\frac{b}{a}\right)\frac{b^2}{2} - \ln\left(\frac{d}{a}\right)\frac{d^2}{2} - \frac{1}{4}(b^2 - d^2)\right\} \end{array}\right], \tag{1.29}$$

where A is defined as

$$A = \frac{d^2 - c^2}{2(2v-1)} - \frac{(b^2 - d^2)^2}{2\{b^2 + d^2(2v-1)\}} - \left(\frac{c^2 - a^2}{2}\right)$$
$$\left\{\frac{2vb^2}{(2v-1)\{b^2 + d^2(2v-1)\}2v\sqrt{2(1-v)}}\right.$$
$$\left(\left(\frac{c}{d}\right)^{-1+\sqrt{2(1-v)}}\left(1 - v + v\sqrt{2(1-v)}\right)\right)$$
$$\left.-\left(\frac{c}{d}\right)^{-1-\sqrt{2(1-v)}}\left(1 - v - v\sqrt{2(1-v)}\right)\right) - \frac{1}{(2v-1)}\right\}$$
$$-\left\{\frac{d^{1+\sqrt{2(1-v)}} - c^{1+\sqrt{2(1-v)}}}{1+\sqrt{2(1-v)}}\right\}\frac{2vb^2}{(2v-1)\{b^2 + d^2(2v-1)\}}$$

$$d^{1-\sqrt{2(1-v)}} \left\{ \frac{1-v+v\sqrt{2(1-v)}}{2v\sqrt{2(1-v)}} \right\}$$

$$+ \left\{ \frac{d^{1-\sqrt{2(1-v)}} - c^{1-\sqrt{2(1-v)}}}{1-\sqrt{2(1-v)}} \right\} \frac{2vb^2}{(2v-1)\{b^2+d^2(2v-1)\}} \qquad (1.30)$$

$$\left\{ \frac{1-v-v\sqrt{2(1-v)}}{d^{-1-\sqrt{2(1-v)}}2v\sqrt{2(1-v)}} \right\}.$$

During the second stage of elastic–plastic deformation, the residual stresses in the plastic zones I and II are given in Eqs. (1.1) and (1.2). The residual stresses in the elastic zone and two outer plastic zones as shown in Fig. 1.1b are given as follows [10]:

Elastic zone, $d \leq r \leq e$:

$$\sigma_r^{el} = \frac{E\alpha}{(2v-1)} T_a + \frac{E\alpha(T_b - T_a)}{2(1-v)\ln\left(\frac{b}{a}\right)} \left[\frac{1}{(2v-1)} \left\{ \ln\left(\frac{d}{a}\right) + \frac{1}{2} - \frac{e^2}{2d^2} \right\} \right.$$
$$\left. + \frac{1}{2} - \frac{e^2}{2r^2} - \ln\left(\frac{b}{a}\right) \left(1 - \frac{a^2}{r^2} \right) \frac{b^2}{b^2 - a^2} \right] \qquad (1.31)$$
$$+ \frac{k_1\sigma_Y}{2v-1} \left(1 + \frac{e^2}{2d^2} \right) + \frac{e^2}{2r^2} k_1\sigma_Y + \frac{E\varepsilon_0}{(1-2v)},$$

$$\sigma_\theta^{el} = \frac{E\alpha}{(2v-1)} T_a + \frac{E\alpha(T_b - T_a)}{2(1-v)\ln\left(\frac{b}{a}\right)} \left[\frac{1}{(2v-1)} \left\{ \ln\left(\frac{d}{a}\right) + \frac{1}{2} - \frac{e^2}{2d^2} \right\} \right.$$
$$\left. + \frac{1}{2} + \frac{e^2}{2r^2} - \ln\left(\frac{b}{a}\right) \left(1 + \frac{a^2}{r^2} \right) \frac{b^2}{b^2 - a^2} \right] \qquad (1.32)$$
$$+ \frac{k_1\sigma_Y}{2v-1} \left(1 + \frac{e^2}{2d^2} \right) - \frac{e^2}{2r^2} k_1\sigma_Y + \frac{E\varepsilon_0}{(1-2v)},$$

$$\sigma_z^{el} = \frac{E\alpha}{(2v-1)} T_a + \frac{E\alpha(T_b - T_a)}{2(1-v)\ln\left(\frac{b}{a}\right)} \left[\frac{2v}{(2v-1)} \left\{ \ln\left(\frac{d}{a}\right) + \frac{1}{2} - \frac{e^2}{2d^2} \right\} \right.$$
$$\left. + 1 - \ln\left(\frac{b}{a}\right) \frac{2b^2}{b^2 - a^2} \right] + \frac{2v}{2v-1} k_1\sigma_Y \left(1 + \frac{e^2}{2d^2} \right) + \frac{E\varepsilon_0}{(1-2v)}. \qquad (1.33)$$

Plastic zone III, $f \leq r \leq b$:

$$\sigma_r^{III} = -k_1\sigma_Y \ln r + C_7 + \frac{E\alpha(T_b - T_a)}{2(1-v)\ln\left(\frac{b}{a}\right)} \left\{ \ln\left(\frac{r}{a}\right) - \ln\left(\frac{b}{a}\right) \left(1 - \frac{a^2}{r^2} \right) \frac{b^2}{b^2 - a^2} \right\},$$
$$(1.34)$$

$$\sigma_\theta^{\text{III}} = -k_1\sigma_Y(1 + \ln r) + C_7 + \frac{E\alpha(T_b - T_a)}{2(1-v)\ln\left(\frac{b}{a}\right)}\left\{1 + \ln\left(\frac{r}{a}\right) - \ln\left(\frac{b}{a}\right)\left(1 + \frac{a^2}{r^2}\right)\frac{b^2}{b^2 - a^2}\right\},$$

$$(1.35)$$

$$\sigma_z^{\text{III}} = -k_1\sigma_Y(1 + \ln r) + C_7 + \frac{E\alpha(T_b - T_a)}{2(1-v)\ln\left(\frac{b}{a}\right)}\left\{1 + 2\ln\left(\frac{r}{a}\right) - \ln\left(\frac{b}{a}\right)\frac{2b^2}{b^2 - a^2}\right\}.$$

$$(1.36)$$

Plastic zone IV, $e \le r \le f$:

$$\sigma_r^{\text{IV}} = \frac{E\alpha}{(2v-1)}T_a + \frac{E\alpha(T_b - T_a)}{2(1-v)\ln\left(\frac{b}{a}\right)}\left[\frac{1}{(2v-1)}\left\{\ln\left(\frac{d}{a}\right) + \frac{1}{2} - \frac{e^2}{2d^2}\right\}\right.$$
$$\left. + \ln\left(\frac{r}{e}\right) - \ln\left(\frac{b}{a}\right)\left(1 - \frac{a^2}{r^2}\right)\frac{b^2}{b^2 - a^2}\right]$$
$$+ k_1\sigma_Y\left\{\frac{1}{2v-1}\left(1 + \frac{e^2}{2d^2}\right) + \frac{1}{2} - \ln\left(\frac{r}{e}\right)\right\} + \frac{E\varepsilon_0}{(1-2v)},$$

$$(1.37)$$

$$\sigma_\theta^{\text{IV}} = \frac{E\alpha}{(2v-1)}T_a + \frac{E\alpha(T_b - T_a)}{2(1-v)\ln\left(\frac{b}{a}\right)}\left[\frac{1}{(2v-1)}\left\{\ln\left(\frac{d}{a}\right) + \frac{1}{2} - \frac{e^2}{2d^2}\right\}\right.$$
$$\left. + 1 + \ln\left(\frac{r}{e}\right) - \ln\left(\frac{b}{a}\right)\left(1 + \frac{a^2}{r^2}\right)\frac{b^2}{b^2 - a^2}\right]$$
$$+ \left\{\frac{1}{2v-1}\left(1 + \frac{e^2}{2d^2}\right) - \frac{1}{2} - \ln\left(\frac{r}{e}\right)\right\}k_1\sigma_Y + \frac{E\varepsilon_0}{(1-2v)},$$

$$(1.38)$$

$$\sigma_z^{\text{IV}} = \frac{E\alpha}{(2v-1)}T_a + \frac{E\alpha(T_b - T_a)}{2(1-v)\ln\left(\frac{b}{a}\right)}\left[\frac{2v}{(2v-1)}\left\{\ln\left(\frac{d}{a}\right) + \frac{1}{2} - \frac{e^2}{2d^2}\right\}\right.$$
$$\left. + 2v\ln\left(\frac{r}{e}\right) + 1 - \ln\left(\frac{b}{a}\right)\frac{2b^2}{b^2 - a^2}\right]$$
$$+ k_1\sigma_Y\left\{\frac{2v}{2v-1}\left(1 + \frac{e^2}{2d^2}\right) - 2v\ln\left(\frac{r}{e}\right)\right\} + \frac{E\varepsilon_0}{(1-2v)}.$$

$$(1.39)$$

In the above equations, e is the interface radius between the elastic zone and plastic zone III and f is the interface radius between the plastic zone III and plastic zone IV. In the second stage of elastic–plastic deformation, radii c, d, e and f are evaluated by using the boundary conditions of vanishing radial stress at the inner and outer radii, $\sigma_\theta^{(\text{plastic zone II})} = \sigma_z^{(\text{plastic zone II})}$ at $r = c$ and $\sigma_\theta^{(\text{plastic zone IV})} = \sigma_z^{(\text{plastic zone IV})}$, at $r = f$ [10]. The various constants involved in the residual stress equations of different zones in the second stage of elastic–plastic deformation are given by

$$C_6 = \frac{E\alpha(T_b - T_a)}{(2v - 1)\ln\left(\frac{b}{a}\right)}\left\{\frac{1 - v - v\sqrt{2(1 - v)}}{d^{-1-\sqrt{2(1-v)}}2v\sqrt{2(1 - v)}}\right\}$$

$$\left\{\ln\left(\frac{d}{a}\right) - \frac{2v + 1}{2v - 1} - \frac{(1 + v)}{1 - v - v\sqrt{2(1 - v)}}\right\}$$

$$- \frac{E\alpha(T_b - T_a)}{2(1 - v)\ln\left(\frac{b}{a}\right)}\left\{\frac{1 - v - v\sqrt{2(1 - v)}}{d^{-1-\sqrt{2(1-v)}}2v\sqrt{2(1 - v)}}\right\} \tag{1.40}$$

$$\left[\frac{1}{(2v - 1)}\left\{\ln\left(\frac{d}{a}\right) + \frac{1}{2} - \frac{e^2}{2d^2}\right\} - \ln\left(\frac{d}{a}\right) + \frac{1}{2} - \frac{e^2}{2d^2}\right]$$

$$- \left\{\frac{1 - v - v\sqrt{2(1 - v)}}{d^{-1-\sqrt{2(1-v)}}2v\sqrt{2(1 - v)}}\right\}\left(\frac{2v}{2v - 1}\right)k_1\sigma_Y\frac{e^2}{2d^2},$$

$$C_5 = -C_6 d^{-2\sqrt{2(1-v)}}\left\{\frac{1 - v + v\sqrt{2(1 - v)}}{1 - v - v\sqrt{2(1 - v)}}\right\}$$

$$- \frac{E\alpha(T_b - T_a)}{(2v - 1)\ln\left(\frac{b}{a}\right)}\frac{(1 + v)}{d^{-1+\sqrt{2(1-v)}}\left\{1 - v - v\sqrt{2(1 - v)}\right\}}, \tag{1.41}$$

$$C_3 = N + \frac{E\varepsilon_0}{(1 - 2v)}, \tag{1.42}$$

where

$$N = C_5 c^{-1+\sqrt{2(1-v)}} + C_6 c^{-1-\sqrt{2(1-v)}} + \frac{k_1\sigma_Y}{(2v - 1)} + \frac{E\alpha T_a}{(2v - 1)}$$

$$+ \frac{E\alpha(T_b - T_a)}{(2v - 1)\ln\left(\frac{b}{a}\right)}\left\{\ln\left(\frac{c}{a}\right) - \frac{2v + 1}{2v - 1}\right\} - k_1\sigma_Y \ln c. \tag{1.43}$$

$$C_7 = M + \frac{E\varepsilon_0}{(1 - 2v)}, \tag{1.44}$$

where

$$M = \frac{E\alpha}{(2v - 1)}T_a + \frac{E\alpha(T_b - T_a)}{2(1 - v)\ln\left(\frac{b}{a}\right)}\left[\frac{1}{(2v - 1)}\left\{\ln\left(\frac{d}{a}\right) + \frac{1}{2} - \frac{e^2}{2d^2}\right\} - \ln\left(\frac{e}{a}\right)\right]$$

$$+ \left\{\frac{1}{2v - 1}\left(1 + \frac{e^2}{2d^2}\right) + \frac{1}{2} - \ln\left(\frac{f}{e}\right) + \ln f\right\}k_1\sigma_Y. \tag{1.45}$$

$$C_9 = -\frac{e^2 \alpha (T_b - T_a)}{2(1-v)\ln\left(\frac{b}{a}\right)}\left(v^2 - 1\right) - e^2\left(1 - v^2\right)\frac{k_1\sigma_Y}{E}. \tag{1.46}$$

$$\begin{aligned}
C_8 &= \frac{f^2 \alpha (T_b - T_a)}{2(1-v)\ln\left(\frac{b}{a}\right)}\left(v^2 - 1\right) + \left(1 - v^2\right)\frac{k_1\sigma_Y}{E}f^2 + C_9 \\
&\quad - \frac{1-2v}{E}\left[\frac{f^2}{2}\left(C_7 - k_1\sigma_Y \ln f\right)\right] - \frac{3-2v}{4E}f^2 k_1\sigma_Y \\
&\quad - \frac{1}{2}f^2 \alpha T_a - \frac{\alpha(T_b - T_a)}{2\ln\left(\frac{b}{a}\right)}f^2\left\{\ln\left(\frac{f}{a}\right) - \frac{3}{2}\right\} + \frac{f^2}{2}\varepsilon_0,
\end{aligned} \tag{1.47}$$

$$\begin{aligned}
\varepsilon_0 &= \frac{k_1\sigma_Y}{BE}\left[
\begin{aligned}
&\ln c\left(\frac{c^2}{2}\right) - \ln a\left(\frac{a^2}{2}\right) + \frac{1}{4}(c^2 - a^2) + \frac{v}{2v-1}(d^2 - c^2) + \frac{v}{2v-1}\left(1 + \frac{e^2}{2d^2}\right)(f^2 - d^2) \\
&- v\left\{\ln\left(\frac{f}{e}\right)f^2 - \frac{1}{2}(f^2 - e^2)\right\} - \frac{1}{4}(b^2 - f^2) - \ln b\left(\frac{b^2}{2}\right) + \ln f\left(\frac{f^2}{2}\right)
\end{aligned}
\right] \\
&\quad + \frac{N}{2BE}(c^2 - a^2) + \frac{C_5}{BE}\left\{\frac{d^{1+\sqrt{2(1-v)}} - c^{1+\sqrt{2(1-v)}}}{1+\sqrt{2(1-v)}}\right\} \\
&\quad + \frac{C_6}{BE}\left\{\frac{d^{1-\sqrt{2(1-v)}} - c^{1-\sqrt{2(1-v)}}}{1-\sqrt{2(1-v)}}\right\} + \frac{\alpha T_a}{2(2v-1)B}(f^2 - c^2) \\
&\quad + \frac{\alpha(T_b - T_a)}{(2v-1)\ln\left(\frac{b}{a}\right)B}\left\{\ln\left(\frac{d}{a}\right)\frac{d^2}{2} - \ln\left(\frac{c}{a}\right)\frac{c^2}{2} - \frac{3}{4}(d^2 - c^2) - \frac{d^2 - c^2}{2v-1}\right\} \\
&\quad + \frac{M}{2BE}(b^2 - f^2) + \frac{\alpha(T_b - T_a)}{2(1-v)\ln\left(\frac{b}{a}\right)B} \\
&\quad \left[
\begin{aligned}
&\frac{v}{(2v-1)}\left\{\ln\left(\frac{d}{a}\right) + \frac{1}{2} - \frac{e^2}{2d^2}\right\}(f^2 - d^2) - \left\{\ln\left(\frac{e}{a}\right)e^2 - \ln\left(\frac{d}{a}\right)d^2 - \frac{1}{2}(e^2 - d^2)\right\} \\
&+ v\left\{\ln\left(\frac{f}{e}\right)f^2 - \frac{1}{2}(f^2 - e^2)\right\} - \left\{\ln\left(\frac{f}{a}\right)f^2 - \ln\left(\frac{e}{a}\right)e^2 - \frac{1}{2}(f^2 - e^2)\right\}
\end{aligned}
\right],
\end{aligned} \tag{1.48}$$

where

$$B = \left\{\frac{b^2 - a^2}{2(2v-1)}\right\}. \tag{1.49}$$

Appendix B: Stress Analysis in the shrink-fitting of cylinders

An outer cylindrical layer of inner radius less than the outer radius b of the inner cylinder by δ, i.e. $(b - \delta)$ is considered, where δ is the shrink-fit allowance. The radius of the outer cylinder is z. A compound cylinder is formed by shrink-fitting the outer cylinder to the inner cylinder. The geometry of the compound cylinder is shown in Fig. 1.2. Due to shrink-fit, the contact pressure p_{sh} is generated between the two cylinders. The contact pressure p_{sh} acts as an external pressure on the inner thermally autofrettaged cylinder and as an internal pressure on the outer cylinder. The stresses in the inner and outer cylinders due to shrink-fit can be obtained as follows.

The radial stress σ_r and hoop stress σ_θ at any point in the wall cross section of a thick-walled cylinder are given by Lame's equations as

$$\sigma_r = A + \frac{B}{r^2}, \tag{1.50}$$

$$\sigma_\theta = A - \frac{B}{r^2}, \tag{1.51}$$

where A and B are constants. Now, for the inner cylinder, the contact pressure p_{sh} due to shrink-fit acts as an external pressure. Hence, we get for the inner cylinder

$$(\sigma_r)|_{r=a} = 0, \tag{1.52}$$

$$(\sigma_r)|_{r=b} = p_{sh}. \tag{1.53}$$

Using Eq. (1.50) in Eqs. (1.52) and (1.53) and then solving, them one can obtain the constants A and B. Substituting A and B in Eqs. (1.50) and (1.51), the resulting stresses in the inner cylinder are obtained as

$$\sigma_r^{sh_i} = -p_{sh} \frac{b^2}{b^2 - a^2} \left(1 - \frac{a^2}{r^2} \right), \tag{1.54}$$

$$\sigma_\theta^{sh_i} = -p_{sh} \frac{b^2}{b^2 - a^2} \left(1 + \frac{a^2}{r^2} \right). \tag{1.55}$$

For the outer cylinder, p_{sh} acts as an internal pressure. Thus, for the outer cylinder

$$(\sigma_r)|_{r=b} = -p_{sh}, \tag{1.56}$$

$$(\sigma_r)|_{r=z} = 0. \tag{1.57}$$

Using Eq. (1.50) in Eqs. (1.56) and (1.57), the constants A and B can be evaluated. The resulting stresses in the outer cylinder are obtained as

$$\sigma_r^{sh_o} = p_{sh} \frac{b^2}{z^2 - b^2} \left(1 - \frac{z^2}{r^2} \right), \tag{1.58}$$

$$\sigma_r^{sh_o} = p_{sh} \frac{b^2}{z^2 - b^2} \left(1 + \frac{z^2}{r^2}\right), \tag{1.59}$$

As the cylinder is open-ended, the axial stress component is zero for the inner as well as for the outer cylinder. The contact pressure p_{sh} can be evaluated as follows.

The radial displacement at any point in the wall of the cylinder can be obtained by using strain–displacement relation along with the generalized Hook's law,

$$\varepsilon_\theta = \frac{u}{r} = \frac{1}{E}(\sigma_\theta - v\sigma_r). \tag{1.60}$$

Evaluating the value of σ_r and σ_θ at the inner wall of the outer cylinder, i.e. at $r = b$ from Eqs. (1.58) and (1.59) and then substituting them in Eq. (60), the radial displacement u at the contact surface of the outer cylinder is obtained as

$$\left(u|_{r=b}\right)_{\text{outer cylinder}} = \frac{p_{sh}b}{E_1}\left(\frac{z^2 + b^2}{z^2 - b^2} + v_1\right), \tag{1.61}$$

where E_1 and v_1 are the Young's modulus of elasticity and Poisson's ratio of the outer cylinder. Similarly, the radial displacement at the contact surface of the inner cylinder can be obtained as

$$\left(u|_{r=b}\right)_{\text{inner cylinder}} = -\frac{p_{sh}b}{E_2}\left(\frac{b^2 + a^2}{b^2 - a^2} - v_2\right), \tag{1.62}$$

where E_2 and v_2 are the Young's modulus of elasticity and Poisson's ratio of the inner cylinder. The total interference δ at the contacting surface can be obtained as

$$\delta = \left(u|_{r=b}\right)_{\text{outer cylinder}} - \left(u|_{r=b}\right)_{\text{inner cylinder}},$$

$$\delta = p_{sh}b\left\{\frac{1}{E_1}\left(\frac{z^2 + b^2}{z^2 - b^2} + v_1\right) + \frac{1}{E_2}\left(\frac{b^2 + a^2}{b^2 - a^2} - v_2\right)\right\}. \tag{1.63}$$

Equation (1.63) provides the value of the contact pressure p_{sh} as

$$p_{sh} = \frac{\left(\frac{\delta}{b}\right)}{\left\{\frac{1}{E_1}\left(\frac{z^2+b^2}{z^2-b^2} + v_1\right) + \frac{1}{E_2}\left(\frac{b^2+a^2}{b^2-a^2} - v_2\right)\right\}}. \tag{1.64}$$

If both the inner and outer cylinder is made of the same material, then $E_1 = E_2 = E$ and $v_1 = v_2 = v$. Thus, Eq. (1.64) reduces to

$$p_{sh} = \frac{\left(\frac{\delta}{b}\right)}{\frac{1}{E}\left\{\left(\frac{z^2 + b^2}{z^2 - b^2}\right) + \left(\frac{b^2 + a^2}{b^2 - a^2}\right)\right\}} . \qquad (1.65)$$

References

1. Abdelsalam, O.R., Sedaghati, R.: Design optimization of compound cylinders subjected to autofrettage and shrink-fitting processes. ASME J. Press. Vessel Technol. **135**, 021209-1–021209-11 (2013)
2. ASME Boiler and Pressure Vessel Code: Rules for Construction of High Pressure Vessels, Section VIII, Division 3, Article KD-4, pp 74–76 (2007)
3. Barsom, J.M., Rolfe, S.T.: Fracture and Fatigue Control in Structures—Applications of Fracture Mechanics. Prentice-Hall, Inc., Englwood Cliffs (1987)
4. Bhatnagar, R.M.: Modelling, validation and design of autofrettage and compound cylinder. Eur. J. Mech. A/Solids **39**, 17–25 (2013)
5. Davidson, T.E., Barton, C.S., Reiner, A.N., Kendall, D.P.: New approach to the autofrettage of high-strength cylinders. Exp. Mech. **2**, 33–40 (1962)
6. Gexia, Y., Hongzhao, L.: An analytical solution of residual stresses for shrink-fit two-layer cylinders after autofrettage based on actual material behavior. ASME J. Press. Vessel Technol. **134**(6), 061209-1–061209-8 (2012)
7. Jacob, L.: La Résistance et L'équilibre Elastique des Tubes Frettés. Mémoire de L'artillerie Navale **1**(5), 43–155 (1907). (in French)
8. Jahed, H., Farshi, B., Karimi, M.: Optimum autofrettage and shrink-fit combination in multi-layer cylinders. ASME J. Press. Vessel Technol. **128**(2), 196–200 (2006)
9. Kamal, S.M.: A theoretical and experimental study of thermal autofrettage process. Ph.D. Thesis, IIT Guwahati, Guwahati, India (2016)
10. Kamal, S.M., Dixit, U.S.: Feasibility study of thermal autofrettage of thick-walled cylinders. ASME J. Press. Vessel Technol. **137**(6), 061207-1–061207-18 (2015)
11. Kamal, S.M., Dixit, U.S.: A comparative study of thermal and hydraulic autofrettage. J. Mech. Sci. Technol. **30**(6), 2483–2496 (2016)
12. Kamal, S.M., Dixit, U.S.: A study on enhancing the performance of thermally autofrettaged cylinder through shrink-fitting. ASME J. Manuf. Sci. Eng. **138**(9), 094501-1–094501-5 (2016b)
13. Kapp, J.A., Brown, B., LaBombard, E.J., Lorenz, H.A.: On the design of high durability high pressure vessels. Proc. ASME PVP Conf. **371**, 85–91 (1998)
14. Rees, D.W.A.: The fatigue life of thick-walled autofrettaged cylinders with closed ends. Fatigue Fract. Eng. Mater. Struct. **14**(1), 51−68 (1991)
15. Mote, J.D., Ching, L.K.W., Knight, R.E., Fay, R.J., Kaplan, M.A.: Explosive Autofrettage of Cannon Barrels, AMMRC CR 70-25, p. 02172. Army Materials and Mechanics Research Center, Watertown (1971)
16. Paris, P.C., Gomez, M.P., Anderson, W.E.: A rational analytic theory of fatigue. Trend Eng. **13**, 9–14 (1961)
17. Parker, A.P., Kendall, D.P.: Residual stresses and lifetimes of tubes subjected to shrink-fit prior to autofrettage. ASME J. Press. Vessel Technol. **125**(3), 282–286 (2003)

18. Perl, M., Aroné, R.: Stress intensity factors for a radial multi-jacketed partially autofrettaged pressurized thick-walled cylinder. ASME J. Press. Vessel Technol. **110**(1988), 147–154 (1988)
19. Rogan, J.: Fatigue strength and mode of fracture of high pressure tubing made from low-alloy high strength steels. In: High Pressure Engineering, I. Mech E., London, UK, pp. 287 − 295 (1975)
20. Sanford, R.J.: Principles of Fracture Mechanics. Prentice hall, Upper Saddle River (2003)
21. Shufen, R., Dixit, U.S.: A finite element method study of combined hydraulic and thermal autofrettage process. ASME J. Press. Vessel Technol. **139**(4), 041204-1–041204-9 (2017)
22. Srinath, L.S.: Advanced Mechanics of Solids. Tata McGraw-Hill, New Delhi (2003)
23. Stacey, A., Webster, G.A.: Determination of residual stress distributions in autofrettaged tubing. Int. J. Press. Vessel Piping **31**, 205–220 (1988)
24. Underwood, J.H.: Stress Intensity Factors for Internally Pressurized Thick-Walled Cylinders. ASTM STP 513, Part 1, pp 59–70 (1972)
25. Zare, H.R., Darijani, H.: A novel autofrettage method for strengthening and design of thick-walled cylinders. Mater. Des. **105**, 366–374 (2016)
26. Zare, H.R., Darijani, H.: Strengthening and design of the linear hardening thick-walled cylinders using the new method of rotational autofrettage. Int. J. Mech. Sci. **124–125**, 1–8 (2017)

Chapter 2
Deformation Behaviour and Fracture Mechanism of Ultrafine-Grained Aluminium Developed by Cryorolling

A. Dhal, S. K. Panigrahi and M. S. Shunmugam

Abstract This chapter highlights the fundamental deformation and fracture mechanism of an engineered ultrafine-grained (UFG) material developed by a combination of cryorolling and short-annealing treatment. The UFG material developed by cryorolling possesses superlative tensile strength. However, the ductility and strain hardening potential of the material is found to be low, reducing its manufacturing capabilities. Controlled post-deformation annealing results in a combination of good strength and ductility. The anisotropic property of the material is also improved after short-term annealing. These properties have been attributed to the unique equiaxed, thermally stable microstructure comprising of high-angled nanometric grains. The various mechanical properties have been experimentally evaluated by performing the tensile test at all three different processing conditions (base, cryorolled, annealed) and the corresponding strain hardening potential, fracture behaviour and anisotropic properties have been systematically investigated. These properties have been correlated with the microstructural features of the material. This has been achieved by mechanical testing and characterisation of the material by employing transmission electron microscopy, fractographic analysis and determination of mechanical anisotropy coefficient (Lankford coefficient). Finally, a case study on the improved microforming abilities of UFG material over coarse-grained material has been presented.

Keywords Ultrafine-grained materials · Cryorolling · Structure–property–manufacturability correlation · Strain hardening potential · Fracture behaviour

A. Dhal · S. K. Panigrahi · M. S. Shunmugam (✉)
Department of Mechanical Engineering, Indian Institute of Technology Madras,
Chennai 600036, India
e-mail: shun@iitm.ac.in

A. Dhal
e-mail: dhal.abhi@gmail.com

S. K. Panigrahi
e-mail: skpanigrahi@iitm.ac.in

© Springer Nature Singapore Pte Ltd. 2019
U. S. Dixit and R. G. Narayanan (eds.), *Strengthening and Joining by Plastic Deformation*, Lecture Notes on Multidisciplinary Industrial Engineering,
https://doi.org/10.1007/978-981-13-0378-4_2

2.1 Introduction

Increase in global warming and carbon dioxide emission is a major worry for many countries all over the world. There is also a rising concern for fuel scarcity as the oil and gas reserves are sinking at an alarming rate. This has led scientists to focus on improving the fuel efficiency of automobiles and aircraft by employing several innovative strategies [1]. One key strategy is to reduce the weight of the automotive and aircraft structures by utilising materials with high strength-to-weight ratio. Modern engineering materials such as composite, titanium alloys, etc., have a high strength-to-weight ratio and are being used abundantly in aerospace and automotive applications. However, these materials are very expensive and difficult to shape by conventional manufacturing processes like forming, machining, etc. As a result, scientists are constantly seeking a solution to enhance the specific strength of traditional engineering materials such as steel, aluminium and copper [2, 3]. Grain refinement is a cost-effective method to improve the specific strength of a wide variety of materials [4, 5].

Ultrafine-grained (UFG) materials are polycrystals which have undergone an extreme level of grain refinement (grain size less than 1 μm). Since past two decades, they have generated a considerable amount of research interest among scientists all over the world. The reason behind this can be attributed to the extraordinary improvement in the mechanical properties (tensile, fatigue strength) and many potential application areas of these materials [6]. Severe plastic deformation (SPD) remains the most popular method to develop such microstructure in the materials. SPD is a "one-step" process to produce UFG materials in bulk form without affecting the geometrical integrity of the material. Unlike, the "two-step" process which involves the development of nanopowder followed by their consolidation, SPD process possess very minimal processing defects such as porosity and impurities [7]. Using different SPD processes, we can develop UFG samples of various sample shapes and size. For example, equal channel angular pressing leads to the develop UFG samples in the form of rods and bars. UFG plates can be developed via friction stir processing and cryorolling. To develop UFG sheets, asymmetric rolling, cryorolling and accumulative roll bonding are suitable SPD techniques [8–10].

Cryorolling is a unique SPD process in which the material is cooled to sub-zero temperatures deformed between two rollers. Because of its sub-zero processing temperature, dynamic recovery is effectively suppressed. This results in a substantially higher dislocation accumulation in the material even with much lesser strain compared to other SPD processes performed at ambient/high temperature [11]. Cryorolling is also simple in its operation, inexpensive and does not require any special tooling. The yield of the process is high, making it suitable for mass production of UFG plates, sheets and foils of various materials. Cryorolled sheets are well suited for many important industries such as aerospace and automotive, where there is a constant requirement of metallic sheets with high strength-to-weight ratio and good workability.

Despite boundless prospectives, UFG materials developed via cryorolling have one major drawback. They have very poor room temperature ductility (as low as 2%) which restricts its forming capabilities. The investigation into the deformation behaviour of various cryorolled materials has been done by many researchers. The primary cause for the restricted room temperature ductility has been attributed to: (i) poor strain hardening potential, (ii) force instability during tension and (iii) shear dominated premature fracture. The root cause of such abnormal deformation behaviour has been investigated by studying the microstructure of cryorolled materials. Transmission electron micrographs of cryorolled aluminium and its alloys reveal dislocation forests or tangles present throughout the material. These dislocation forests consist of numerous amount of dislocations randomly intertwined with each other and lack any definite cellular structure. As a result, the microstructure is supersaturated with dislocations and does not have the capability to absorb any further dislocation, thus restricting its strain hardening capabilities. Apart from this, the onset of non-homogenous and localised deformation via necking also occurs at very low strain level in such materials [12]. The low ductility and insufficient strain hardening ability of UFG materials have limited their practical application and, therefore, methods to improve their strain hardening ability are being given due importance by several researchers.

Annealing is an effective method to remove excessive dislocation from the material microstructure and make the material more ductile. During annealing, first, the randomly crowded dislocations get reordered by: (i) annihilation of an excessive amount of dislocation (recovery) and (ii) rearrangement of the remaining dislocations into definite cellular structures (polygonization). In this way, the overall dislocation density of the material is reduced, and new stress-free grains are recrystallized. The newly recrystallized grain boundaries act as nucleation sites for dislocations formed during tensile deformation. However, further annealing of the same material leads to diffusion assisted microstructural phenomena such as grain growth, rotation and coalescence. This results in an increase in the average grain size of the material and relative misorientation between the grains. For cryorolled materials, due to the high energy state of the processed microstructure, annealing leads to an accelerated rate of recovery, recrystallisation and even premature grain growth. This results in very low thermal stability of the UFG microstructure. However, this is not the case for heat treatable alloys due to the influence of the second phase precipitates. These precipitates are first dissolved in the solid solution before the cryorolling process. Precipitates still remain trapped in the solute matrix as dynamic precipitation is suppressed due to the sub-zero processing temperature during cryorolling. Only after annealing, these precipitates emerge out from the matrix and predominately occupy the grain boundaries. They exert a back pressure on the movement of grain boundaries thereby retarding the rate of grain growth. This phenomenon is termed as Zener pinning and the exerted back pressure depends on the ratio of volume fraction to the mean diameter of the particles [13]. Therefore, a controlled short-term annealing can be employed to reduce the dislocation density in the cryorolled material without much increase in its grain size.

The microstructural modification after controlled annealing leads to substantial improvement in the total ductility and strain hardening potential of the materials. They also influence the anisotropic properties (Lankford coefficient) and fracture mechanism of the cryorolled material. In this chapter, a popularly used commercially pure aluminium alloy (AA1070) has been cryorolled and then annealed to produce UFG microstructure. The influence of cryorolling and annealing on the deformation behaviour and fracture mechanism of the material has been investigated. Finally, a case study on the microforming capabilities of the developed UFG aluminium sheets has been presented.

2.2   Development of a Unique UFG Microstructure via Cryorolling and Annealing

For the present investigation, a commercially pure aluminium alloy (AA1070) has been used. AA1070 contains 99.7% of Al, 0.25% of Fe, 0.2% of Si and traces of Zn, Cu, Mg, V, Ti. This alloy is widely used as a construction and packaging material and for making communication wires. For all these applications, apart from the functional properties, a combination of good mechanical strength and formability is necessary. Therefore, this alloy serves as an ideal material for SPD processing via cryorolling and its successive annealing.

The samples were procured in the form of wrought plates of thickness 10 mm. These plates were subsequently cut down to small square samples of dimension 40 mm × 40 mm × 10 mm. Before processing for cryorolling, the square samples were thoroughly annealed at 400 °C for 2 h, followed by immediate quenching in cold water. This was done to coarsen the grain and make the material sufficiently soft for easy cryorolling processing. This represents the base condition of the material and its microstructure determined by using electron backscatter diffraction (EBSD) technique is shown in Fig. 2.1. The average grain size at base condition was found to be around 31 μm. The material has random texture and does not possess any signature of deformation. The cryorolling was done by using a 2-high rolling mill equipped with rollers of 110 mm and operating at 8 rpm. The base sample was first dipped in a liquid nitrogen bath for 15 min. It was subsequently passed through the two rollers and after each pass, it was re-dipped in the liquid nitrogen for 10 more minutes. This process was repeated for multiple cycles until the required strain (true strain = 3.6) was achieved. The microstructural study of the cryorolled material was also done by the means of transmission electron microscope (TEM).

The TEM micrograph (Fig. 2.2) reveals the intricate details regarding the morphology, orientation and size of the AA1070 after cryorolling. Analysis of the micrograph suggests that the cryorolled AA1070 microstructure consists of elongated cell blocks oriented in a direction nearly 45° to the rolling direction. There is a substantial amount of stored dislocation density in the microstructure and 77%

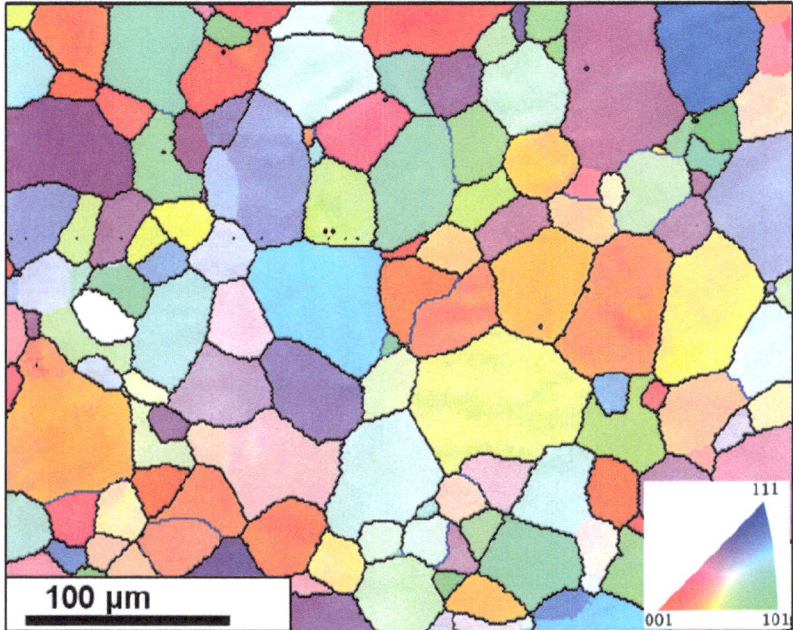

Fig. 2.1 EBSD map showing the microstructure of the base AA1070

grains have low-angled grain boundaries. These dislocations mostly reside in the dense dislocation walls which demark the cell blocks. The mean grain size was found to be nearly 788 nm with a scatter of 392 nm.

To relieve the excessive stored dislocations and to obtain a stress free, equiaxed UFG microstructure, the cryorolled sheet was annealed at 150 °C for 30 min in an enclosed furnace followed by immediate quenching in cold water. The choice of the specific temperature and annealing cycle is based on the information extracted from the authors' previous work which contains the thermal stability analysis of various aluminium alloys after cryorolling [13]. At this temperature, the microstructure contains a significant amount of high-angled ultrafine grains ($\sim 57\%$). This is revealed in the corresponding TEM micrograph (Fig. 2.3), in which the grain morphology appears partially elongated and still carries the signature of rolling even after some heat treatment. It also shows the presence of high-angled UFG microstructure with a significant number of dislocation free grains. The influence of annealing on the grain size is very minimal as the average size just increases to 915 nm, a mere 17% rise from the cryorolled state. A significant dip in the dislocation density could be noted by observing the TEM micrograph.

Quantification of dislocation density (ρ) was done by using a modified form of Williamson–Hall equation [14]:

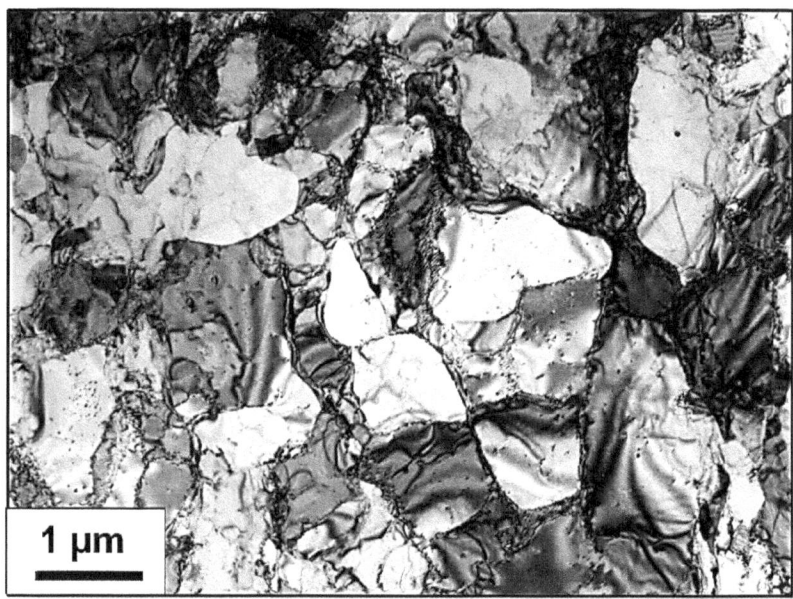

Fig. 2.2 TEM micrograph of AA1070 after cryorolling showing very fine cell blocks structures delineated by dense dislocation walls

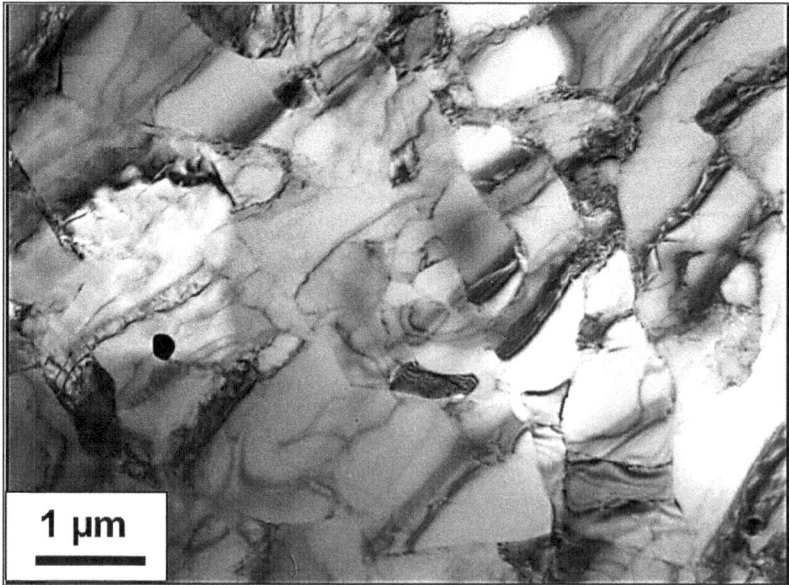

Fig. 2.3 TEM micrograph of cryorolled AA1070 after annealing at 150 °C for 30 min

$$\rho = \frac{2\sqrt{3}}{Db} \left(\frac{B\cos\theta - \lambda K/D}{4\sin\theta} \right) \tag{2.1}$$

where b is the Burger's vector, θ_b is the Bragg's angle, D is the domain/crystalline size, B is the value of full width at half maximum at θ_b, K is a material constant (0.9 for FCC materials [14]) and λ is the wavelength of the X-ray (0.15418 nm for CuK$_\alpha$ radiation). The magnitude of Burgers' vector is determined by using the expression: $||b|| = a/2\sqrt{h^2 + k^2 + l^2}$, where a is the unit cell edge length of the crystal, whose value is 0.404 for pure aluminium [14]. To determine the values of D and B, X-ray diffraction analysis of the {111} Al peak was done for all three processing conditions of AA1070. The influence of background noise and the instrumental broadening effect was removed by calibrating it with respect to a standard specimen. The XRD plot and the trend of dislocation density measured using Eq. 2.1 for all three different processing conditions have been illustrated in Fig. 2.4a. The quantitative measurement of the grain size for all three processing stages of AA1070 was done by analysing the corresponding EBSD/TEM micrographs using linear intercept method. For this, at least 50 individual grains were measured using EBSD/TEM images captured at various sites. The variation of the average grain size and scatter for all three processed conditions has been presented in Fig. 2.4b.

As expected, the dislocation density increases significantly after cryorolling (75% rise). This can be seen in the TEM micrographs (Fig. 2.2) which evidently show dense dislocation walls surrounding the fine rhombic cells. After annealing at 150 °C for 30 min there is a substantial dip in the dislocation density due to accelerated static recovery which has resulted in annihilation and reorganisation of the stored dislocations. This can be envisioned in the corresponding TEM images, which shows microstructure with much less dislocation in the grain boundaries and interiors (Fig. 2.3). The grain size variation remains very minimal even after annealing (Fig. 2.4b). This proves that annealing the cryorolled material at 150 °C for 30 min is sufficient enough to activate static recovery, resulting in dislocation annihilation but is not high enough for recrystallization of new grains and significant growth of the existing grains. This stabilisation of the initial microstructure has been attributed to the rhombic morphology of the cell blocks and significant dynamic recovery during cryorolling, which reduces the driving force for recrystallisation and grain growth.

2.3 Deformation Behaviour of the UFG Material

In this section, the deformation behaviour of the UFG material has been analysed and compared with that of the base and cryorolled conditions. The tensile stress–strain curves have been obtained at all three processing conditions for analysing the

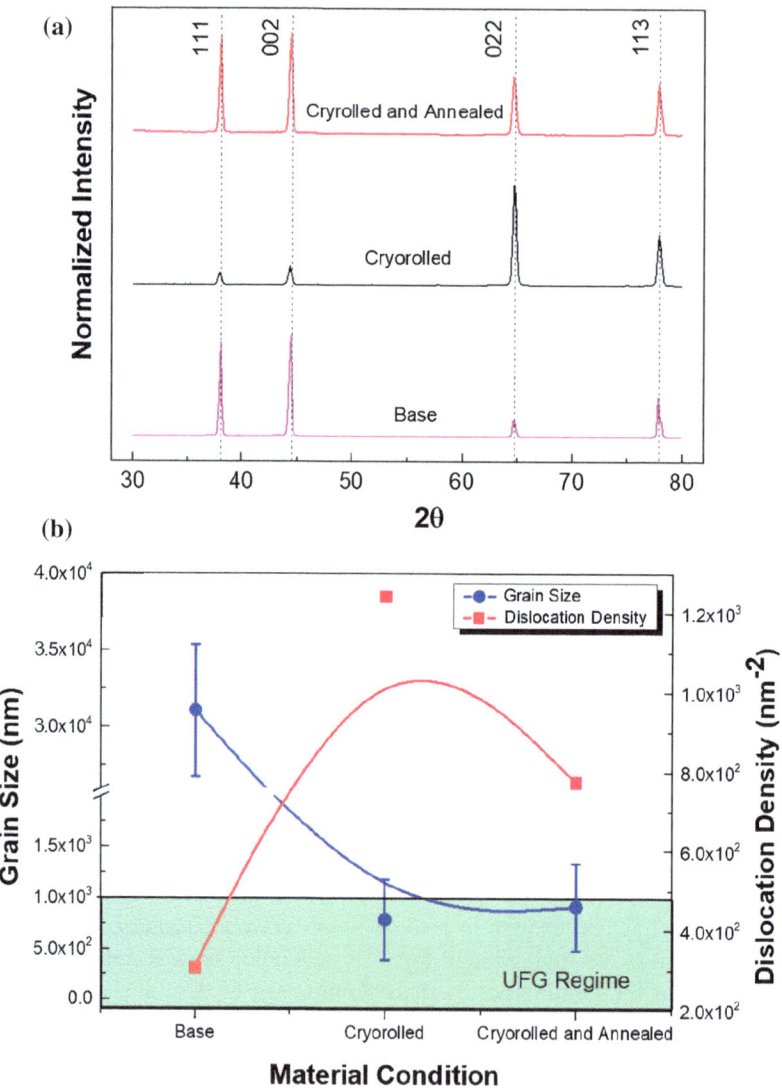

Fig. 2.4 **a** X-ray diffraction plot and **b** dislocation density and grain size variation of AA1070 at three different processing conditions

tensile strength, uniform ductility and strain hardening exponent of the material. These results have been correlated to the corresponding microstructural features of each material.

2.3.1   Strengthening Behaviour of the UFG Material

The tensile properties of the AA1070 were captured at three conditions: (i) base condition (i.e. after pre-SPD heat treatment), (ii) cryorolled condition and (iii) cryorolled + annealed (150 °C, 30 min) condition. For the rolled samples, the gauge length was along the rolling direction of the sheet. The engineering stress–strain curves obtained after the investigation have been shown in Fig. 2.5. The influence of cryorolling can be easily realised by analysing the trend in the curves. Due to cryorolling, there has been an extraordinary increase in the ultimate tensile strength (UTS) of the material. The strength of the cryorolled material (UTS = 187 MPa) increases by a staggering 92% with respect to the base condition (UTS = 95 MPa). After annealing at 150 °C for 30 min, the UTS of the material decreases to 128 MPa, which is a 31% dip compared to the cryorolled material, but still it is 34% greater than the base material.

In the present case, there are two contributing factors responsible for the tensile strength of the materials: (i) dislocation accumulated in form of clusters and tangles, (ii) grain refinement contributing to the Hall–Petch strengthening of the material. In case of cryorolling, the contribution of strength due to dislocation strengthening is much more significant due to heavy accumulation of dislocation in the material owing to the sub-zero deformation temperature. This results in suppression of dynamic recovery, leading to the intertwining of dislocations in the microstructures of the material (evident from clusters and tangles are seen in the TEM micrograph).

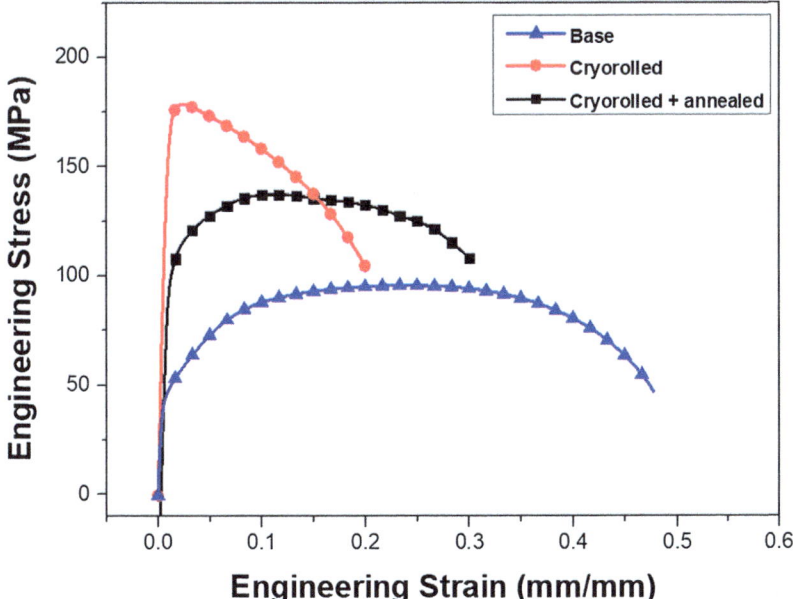

Fig. 2.5 Engineering stress–strain plot of the AA1070 at three different processing conditions

During tensile testing of the cryorolled sample, additional dislocations are generated which interact rapidly with the pre-existing dislocation clusters and tangles. Therefore, greater energy is required to overcome the obstruction provided by the pre-existing dislocations and the stress required to deform the material increases. The decrease in UTS for the annealed sample can be attributed to the significant dip in the dislocation density of the material due to annealing. Annealing activates static recovery within the material resulting in the annihilation of excessive dislocations, mostly stored in the form of unstable tangles or clusters.

Therefore, the contribution of the tensile strength of the material due to dislocation strengthening effect (σ_d) is a direct function of the dislocation density (ρ) and is mathematically represented by means of the following equation:

$$\sigma_d = M\alpha Gb\sqrt{\rho} \qquad (2.2)$$

In this equation, M is the Taylor factor, b is the magnitude of Burgers vector, G is the shear modulus of the material. Literature suggest the value of M for pure aluminium subjected to a true strain of 3.6 is 3.15 and the values of b and G for pure aluminium are 3.5 Å and 26.5 GPa respectively [14].

The second factor contributing to the strength of the cryorolled material is due to the presence of submicrometric-sized UFG microstructure compared to its coarse-grained microstructure in the base condition. These UFG materials have rhombic-shaped grains, which have high relative misorientation angles. They contain non-ordered dislocation tangles and clusters in the grain interiors. Such rhombic grains have formed due to the intersection of thin cell blocks orthogonally aligned to one another. These features are distinctly visible in TEM micrographs of the cryorolled AA1070 (Fig. 2.2).

The reason for the formation of cell block can be attributed to the high stacking fault energy of the AA1070, which activates poly-slip deformation mechanism resulting in the formation of distinct deformation bands/zones each corresponding to a specific slip system. These bands are delineated from each other by well-defined boundaries densely packed with fine dislocation substructures. Grain boundaries are known to be the perfect sites for dislocation nucleation and accumulation during any form of tensile loading. This strengthening phenomenon is called Hall–Petch strengthening effect (σ_{gb}), which correlates the material strength to the average grain size (d) of material [15] in the following manner:

$$\sigma_{gb} = \sigma_0 + \frac{k}{\sqrt{d}} \qquad (2.3)$$

In Eq. 2.3, σ_o is considered as the frictional stress which resists the motion of glide dislocations and k is the grain boundary strengthening coefficient, also known as Hall–Petch slope which denotes the resistance provided by the grain boundary against slip transfer. Both constants depend on the material type and for aluminium, the reported values of σ_o and k are 9.8 MPa and 0.06 $(MPa)m^{1/2}$ respectively [15].

Using Eqs. 2.2 and 2.3 and the values of ρ and d (as obtained from Fig. 2.4), the strength contributed by the dislocations (σ_d), strength contributed by the grain boundaries (σ_{gb}) and the total tensile strength (σ_t) for all three processing conditions have been calculated. In Fig. 2.6, the trend of the theoretical strength value has been plotted along with the experimental strength values (UTS) as obtained from the stress–strain curves for all three processing conditions. The trends of both experimental and theoretical values correlate well with one another.

Since the material is strengthened by means of cryorolling, severe anisotropy in the distribution of tensile strength is expected from the sheet. This can be clearly seen in Fig. 2.7, where the variation of UTS of the AA1070 sheet along three different directions (along with the rolling direction, transverse direction and along 45° to rolling direction) for all three processing conditions of AA1070 have been plotted. For the base sample, the anisotropy in tensile strength is minimal resulting in a flat curve. It is a predictable result due to the homogenous and equiaxed microstructure of the base material. After cryorolling, the strength along the rolling direction is found to be the highest followed by the direction along 45° to rolling direction and finally the transverse direction. At cryorolled state, the predominant contributor to the strength of the material is the accumulated dislocations. During cryorolling, dislocation motion within the sample is very minimal, thereby resulting in greater dislocation accumulation along the rolling direction and minimal accumulation along the transverse direction. Interestingly, for the cryorolled sample annealed at 150 °C, the strength of the material is minimal along 45° to the rolling direction and almost equally maximum along both rolling and transverse directions. Since static recovery has resulted in the annihilation of excessive dislocation in the

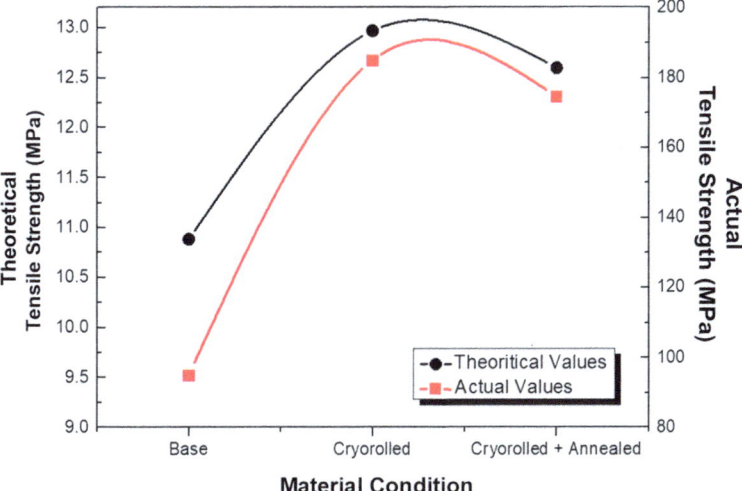

Fig. 2.6 Correlation between theoretical and experimental values of ultimate tensile strength for all three processing conditions of AA1070

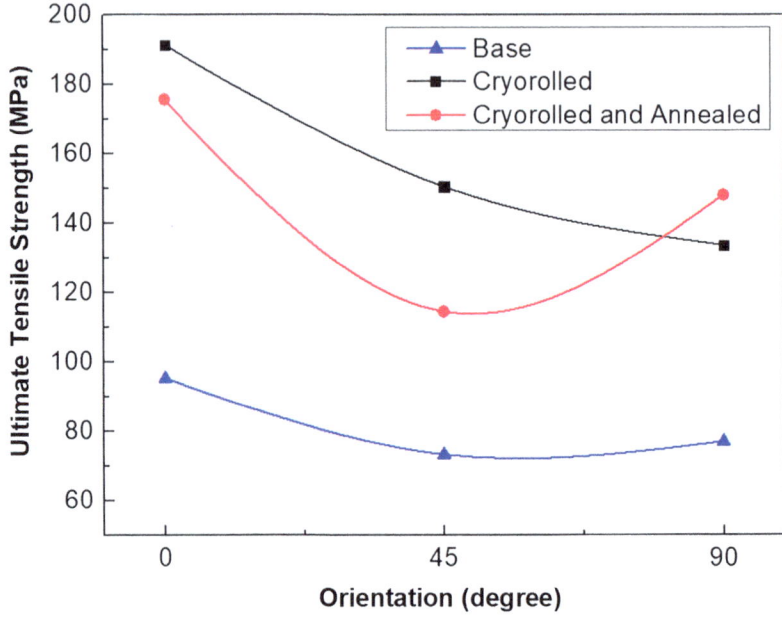

Fig. 2.7 Anisotropy in the distribution of tensile strength for AA1070 processed in all three conditions

material, the observed trend can be attributed to the orientation and morphology of the grain and grain boundaries. Due to poly-slip deformation mechanism prevalent during cryorolling, the microstructure is composed of grains oriented in a crisscross manner along 45° and 135° to the rolling direction. This makes the deformation of the material along 45° direction less strenuous resulting in a decrease in UTS.

2.3.2 Ductility and Strain Hardening Potential of the UFG Material

While cryorolling improves the mechanical strength of the material, on the other hand, the total ductility of the material is sacrificed. This can be seen in the stress–strain curves, where a significant drop in the uniform ductility of the material has been observed after cryorolling (from 35 to 3%). It fits the general notion which states that strength and ductility are mutually exclusive and an increase in either one of the factor leads to the decrease in other. The reason behind this can again be corroborated by the microstructural changes during cryorolling and post-cryorolling annealing. The low total ductility as observed in the cryorolled material may be due the following reasons: (i) the dislocation saturated microstructure obtained in cryorolled material have less dislocation accumulation capability which limits its ductility, (ii) rapid interaction with the pre-existing dislocations causes increased stress concentration and plastic instability in the cryorolled materials resulting localised deformation, which leads to premature necking and subsequent failure of the material.

Annealing of the cryorolled sheets at 150 °C for 30 min leads to improvement in the ductility (uniform ductility = 15%) combined with a slight decrease in strength. Either way, a good combination of strength and ductility has been obtained at this condition and the material processed at this condition looks promising for application in metal forming industries to develop high strength components. The improvement in ductility after annealing can be attributed to the presence of dislocation free UFG microstructure, which has a large number of dislocation nucleation sites. The dislocations can propagate with relative ease without much obstruction due to dislocation free grain boundaries. The onset of necking is also more gradual in case of the annealed sample compared to the cryorolled samples signifying less plastic instability in the annealed samples. This results in a significant improvement of both uniform ductility and total ductility of the material after annealing.

The tensile curves have two distinct plastic deformation regions: (i) uniform deformation zone, where material undergoes strain hardening and (ii) non-uniform deformation zone, where material undergoes localised deformation (necking). The second zone is a precursor to fracture and, in this zone, the material undergoes plastic instability. This usually begins after the material gets strain hardened to the maximum limit (UTS). The uniform deformation zone or the strain hardening zone contributes to the uniform ductility or true ductility of the material and is an essential requirement for sheet metal forming industries. In the present case, the stress–strain curves (Fig. 2.5) reveal a significant change in the strain hardening potential of material after cryorolling. The strain hardening potential of a material is generally quantified by means of strain hardening exponent (m). The strain hardening exponent can be calculated by a modified version of the Crussard–Jaoul equation, which itself is a logarithmic alteration of classical Swift stress–strain relationship [12] and can be expressed as the following equation:

$$\ln\left(\frac{\partial \sigma}{\partial \varepsilon}\right) = (1 - m)\ln \sigma - \ln(Cm) \qquad (2.4)$$

In this equation, C is a material coefficient. The above equation represents a straight line and the value of strain hardening exponent (m) can be calculated by finding the slope of the line. The values of total ductility and strain hardening exponent for AA1070 at all three processing conditions (base, cryorolled and annealed) have been presented in Fig. 2.8.

It can be clearly seen from Fig. 2.8, the value of total ductility and strain hardening exponent for cryorolled AA1070 is the minimum among all three material conditions. The first reason behind this behaviour is the early necking in the material which is caused by a rapid rise in the stress concentration sites due to the resistance provided by the pre-existing dislocation jogs (formed during

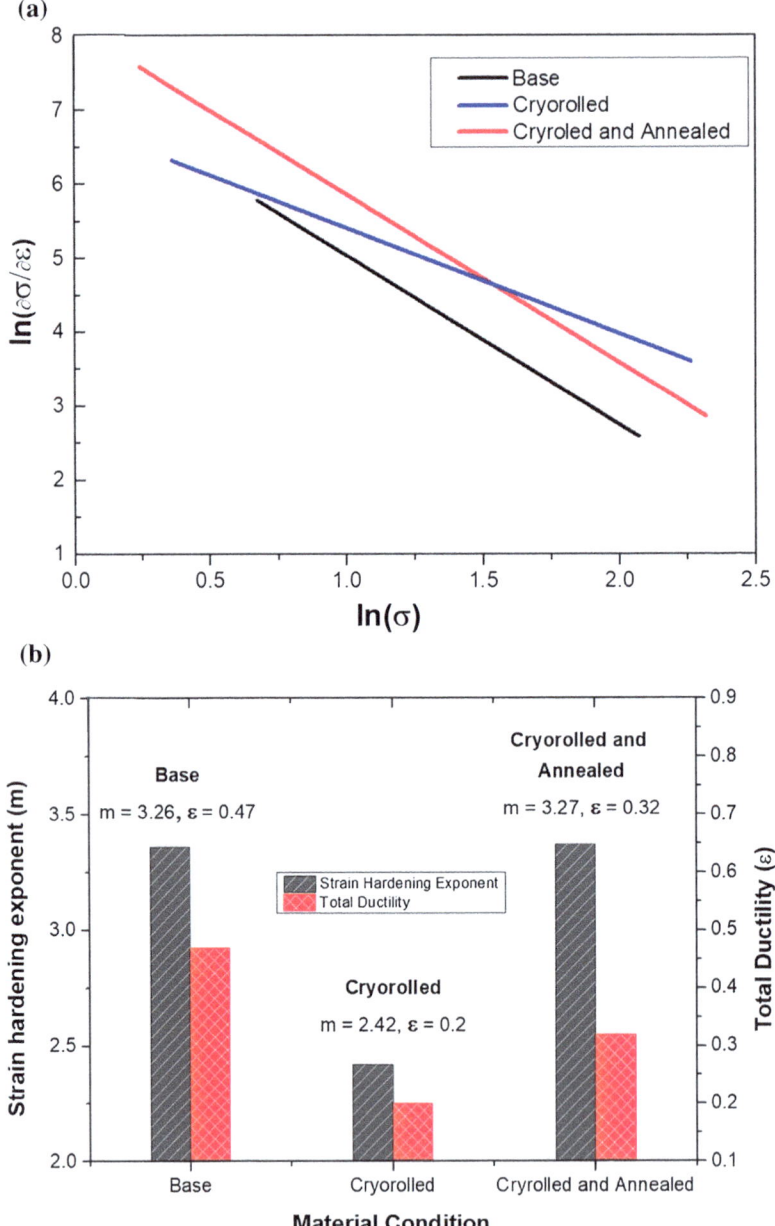

Fig. 2.8 **a** Double-logarithmic plot representing the modified Crussard–Jaoul analysis and **b** variation of strain hardening exponent for three different processing conditions

cryorolling) to the new dislocations during tensile loading. The second reason is the abundant presence of triple junction formed due to grain refinement which also contributes towards rapid stress concentration and early necking of the material. Upon annealing, excessive dislocation densities are removed resulting in a much smoother dislocation movement along the grain boundaries. However, the dislocations are still obstructed by the triple junctions. Therefore, the strain-hardening exponent and uniform ductility are improved but remain lower than that of the base material. In case of base material, dislocations although get accumulated in much lesser amount than UFG material (resulting in lower tensile strength), they move along the grain boundary more easily resulting in a good degree of strain hardening.

2.3.3 Mechanical Anisotropy of the UFG Material

Apart from good ductility, a favourable anisotropy is of utmost importance for good formability of a material. Mechanical anisotropy is usually denoted in terms of two parameters: (i) average anisotropic coefficient or, Lankford coefficient (R_{avg}) and planar anisotropic coefficient (ΔR). Both these factors play an important role in determining the quality of the formed component. R_{avg} is the material's ability to resist thinning when it is subjected to uniform tension. High R_{avg} value is desirable, as it prevents tearing of sheet during deformation. Similarly, the value of ΔR decides the tendency of the material to exhibit earing. For minimal earing, the value of ΔR should be as close as possible to zero. To determine the values of both anisotropic coefficients, tensile samples were cut in three directions: one along rolling direction, one along the transverse direction and final one along 45° to the rolling direction. These samples were subjected to an equal amount of uniform tensile straining. The values of R_{avg} can be calculated as

$$R_{\text{avg}} = \frac{R_0 + 2R_{45} + R_{90}}{4} \tag{2.5}$$

R_0, R_{90} and R_{45} are the mechanical anisotropy along rolling direction, transverse direction and along 45° to the rolling direction. These values can be calculated by using the following equation:

$$\{R\}_{0,45,90} = \left\{ \frac{\log(w_o/w_f)}{\log(l_f w_f/l_o w_o)} \right\} \tag{2.6}$$

where w_o, l_o and w_f, l_f are the length and width of the sample before and after the uniform straining.

Similarly, the value of ΔR can be calculated by

$$\Delta R = \frac{R_0 - 2R_{45} + R_{90}}{2} \tag{2.7}$$

The values of ΔR and R_{avg} for three processing conditions (base, cryorolled and annealed) have been illustrated in Fig. 2.9.

An interesting trend has been observed regarding the anisotropic behaviour of the material processed at the various conditions. The base material, despite its superior ductility, does not possess favourable anisotropic properties. It has lowest R_{avg} value and highest deviation of ΔR from zero. This suggests that the material at base condition is susceptible to thinning, tearing and earing during deformation making it unfit for sheet metal forming applications. The reason for this can be attributed to the microstructure which has more equiaxed grain morphology (Fig. 2.1) and possesses a random crystallographic texture. This leads to equal strain distribution along the plane as well as along the thickness direction of the sheet. Due to this, a substantial amount of material deformation (thinning) occurs along the thickness direction of the sheet. This results in a decrease in the normal as well as the planar anisotropy of the material. Due to poor strain hardening ability

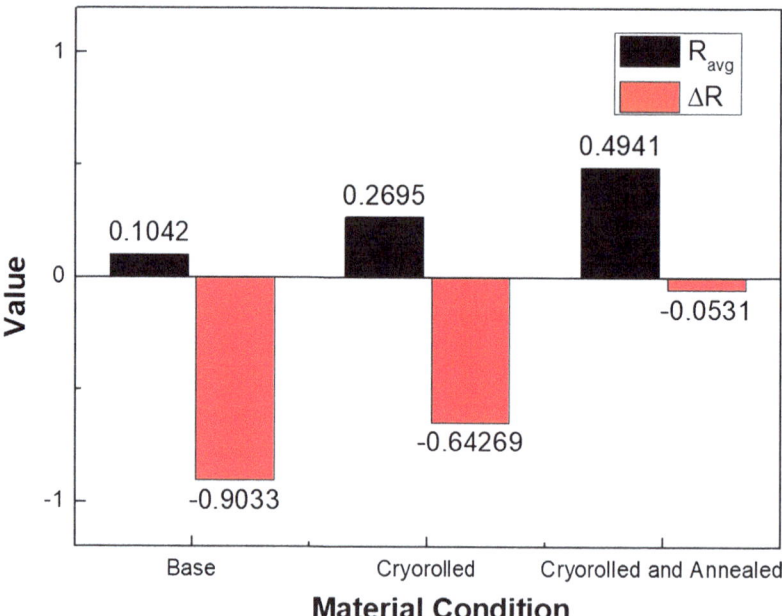

Fig. 2.9 Variation of normal and planar anisotropy values for AA1070 processed at three different conditions

and premature necking (due to the presence of a large amount of dislocations) in the cryorolled sheet, the material thickness reduces down very quickly, resulting in inferior plastic anisotropy. The material becomes more favourably anisotropic as it is subsequently annealed after cryorolling. The orientation of grains along 45° to the rolling direction makes the plastic anisotropy more favourable.

2.4 Fracture Mechanism of the UFG Materials

Fractography analysis was done by post-mortem analysis of the failed samples under a high-resolution scanning electron microscope for all three material processing conditions (base, cryorolled and annealed). The morphology of the failed surface reveals greater details regarding the mode of fracture. For base condition, the fractography reveals the presence of several coarse dimples with size varying from 15 to 28 μm (Fig. 2.10a). Interestingly, the size of the dimples is not very far from the actual grain size of the material in this condition. This shows that grain boundaries being the weak link in the microstructure serve as the fracture initiation sites in the material. These fractures arise due to the coagulation of very small micro-dimples, which have formed due to stress concentration during post necking deformation. These fracture lines eventually propagate along the grain boundaries resulting in the transgranular fracture. This mode of fracture is mostly seen in very ductile material and hence the present observation corroborates well with the large ductility and strain hardening ability of the coarse-grained base material.

On the other hand, the fractograph of cryorolled aluminium shows entirely different morphology (Fig. 2.10b). It is mainly composed of a large number of primary as well as secondary dimples distributed all over the fractured surface. These dimples have an average size of 6 μm with sharp and flat facets and they are slightly elongated in the shear direction. This clearly points out towards a shear-dominated fracture mechanism. Shear-dominated fracture is common in heavily cold-worked materials due to an abnormal rate of plastic instability. This causes the material to shear and separate leaving behind a trail of elongated and shallow dimples.

Annealing leads to decrease in dislocation density thereby reducing the propensity of plastic instability and rapid fracture in the material. However, the presence of grains oriented along 45° and 135°, also favours in a more shear-dominated fracture. The fractography reveals that the fracture lines and facets are also oriented in a similar manner for this material (Fig. 2.10c). However, there is also an abundance of micro-dimples, indicating the occurrence of combined shear and ductile modes of fracture in the material.

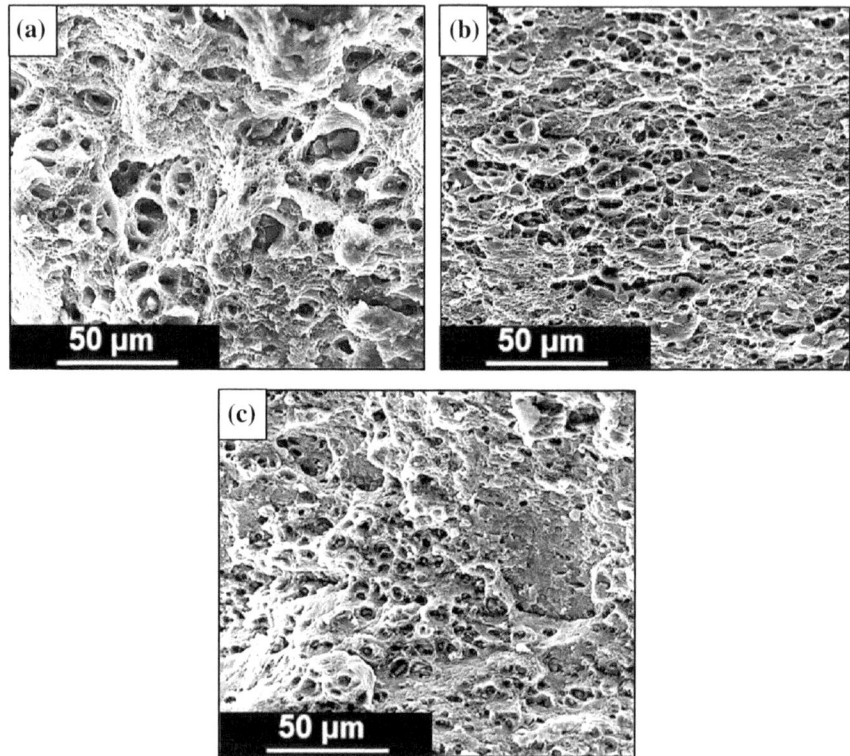

Fig. 2.10 SEM fractography corresponding to AA1070 processed at three different conditions: **a** base, **b** cryorolled and **c** cryorolled and annealed

2.5 Case Study–Microforming Capabilities of the Developed UFG Sheets

The UFG material developed by the above processing route is expected to have good microforming capabilities. The following section compares the microforming capabilities of both the UFG and coarse-grained materials and corroborates this behaviour to their mechanical and microstructural properties.

2.5.1 Size Effect and Its Possible Remedies by Utilising UFG Microstructure

Microforming is the production of parts or structures with at least two dimensions in sub-millimetric level. There are many advantages of microforming over other micro-manufacturing techniques (photolithography, micromachining) such as mass production capabilities, excellent dimensional accuracy and surface finish of the product, near-net-shape production, low energy consumption, stable and consistent product quality, maximum material utilisation and improved mechanical property in the formed components [16, 17]. Despite the multiple advantages of microforming, it has not been popularly translated into industrial application. It is attributed to the lack of substantial knowledge base for microforming. Although metal forming in macro-scale is an age-old manufacturing process, concepts of microforming cannot leverage on the information available for macro forming [18].

The transfer of knowledge from macro to microdomain has been hindered by a phenomenon called 'size effect'. In the context of microforming, size effect occurs when the product size of the formed component becomes comparable with the grain size. This leads to abnormal material behaviours such as (i) decrease in flow stress [19], (ii) anisotropic deformation behaviour [20], (iii) premature fracture [21], (iv) decrease of geometrical accuracy [22] and (v) increase in process scatter [22]. These abnormalities are witnessed due to a decrease in the number of grains in the thickness direction, resulting in a deformation behaviour approaching that of single crystal materials. Using UFG material instead of conventional coarse-grained material is one of the strategies which can be employed to counter the size effect-related abnormalities. UFG microstructure can retain the polycrystallinity nature of the material even if the product size is in few microns.

2.5.2 Correlation of the Mechanical and Anisotropic Properties with the Microformability

To compare the quality of the microformed components, micro deep drawing was performed using both UFG and coarse-grained materials. The results were correlated with their tensile and anisotropic properties. For carrying out micro deep drawing process, a custom designed, the high-precision microforming tool is used. The forming process has been carried out with a low ram speed of 0.3 mm/min until fracture of the materials. After forming, the components have been examined under a stereo-microscope (Zeiss CS2000). The macrographs of the formed-failed components have been shown in Fig. 2.11.

The height of the failed micro-cup for UFG and coarse-grained material was found to be 0.9 and 2.1 mm respectively. The greater drawn height of coarse-grained material can be attributed to its higher ductility. Despite its higher ductility, quality of micro-cup obtained in the coarse-grained material is considerably inferior as compared to the one obtained after drawing UFG material. The coarse-grained material exhibits a significantly larger fracture line at the bottom end of the cup and more earing at the top edges of the cup.

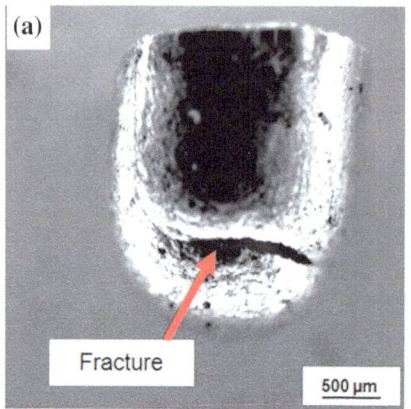 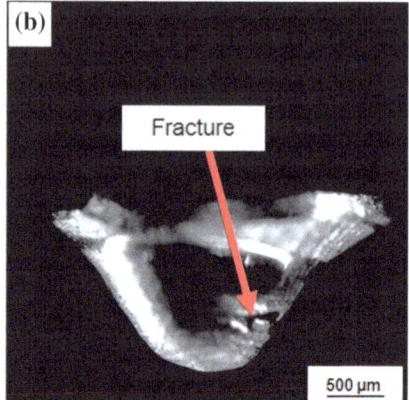

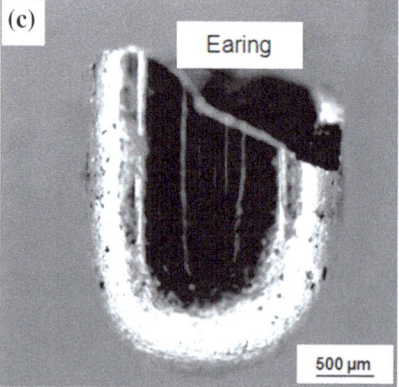

Fig. 2.11 Macrographs of the failed micro-cups obtained by drawing: **a** coarse-grained material (base condition), **b** UFG material (cryorolled + annealed), **c** earing in coarse-grained material [23]

The microforming behaviour of the coarse-grained material can be attributed to the low R_{avg} value, which leads to less resistance to thinning and tearing. As a result of this, the thickness of cup wall is reduced significantly in the coarse-grained material, leading to a prominent fracture line at the cup bottom. The highly negative value of ΔR of coarse-grained material promotes the formation of earing. The influence of size effect is also more prominent in case of coarse-grained material, which contributes to (a) decrease in flow strength during forming, (b) anisotropic deformation mode and (c) premature failure. Therefore, a combination of all these factors could lead to poor quality of the micro-component produced from coarse-grained material compared to UFG material.

2.6   Conclusion

Cryorolling results in extraordinary improvement of strength (92% improvement in UTS compared to base condition). However, the ductility and strain hardening potential of the material is adversely affected. The anisotropic parameters are also not favourable, resulting in the poor microformability. Controlled annealing of the cryorolled material at 150 °C for a duration of 30 min results in a good combination of strength and ductility. This leads to a significant improvement in the strain hardening potential and makes the mechanical anisotropy favourable for sheet metal forming. The UFG microstructure obtained after annealing also tackles the size effect-related abnormalities and improves microforming characteristics of AA1070.

References

1. Fontaras, G., Zacharof, N.G., Ciuffo, B.: Fuel consumption and CO_2 emissions from passenger cars in Europe—laboratory versus real-world emissions. Prog. Energy Combust. Sci. **60**, 97–131 (2017)
2. Sah, S.K., Bawase, M.A., Saraf, M.R.: Light-weight materials and their automotive applications. *SAE Int.* (2014)
3. McAuley, J.W.: Global sustainability and key needs in future automotive design. Environ. Sci. Technol. **37**(23), 5414–5416 (2003)
4. Mohanty, P.S., Gruzleski, J.E.: Mechanism of grain refinement in aluminium. Acta Metall. Mater. **43**(5), 2001–2012 (1995)
5. St. John, D.H., Qian, M., Easton, M.A., Cao, P., Hildebrand, Z.: Grain refinement of magnesium alloys. Metall. Mater. Trans. A **36**(7), 1669–1679 (2005)
6. Valiev, R.: Nanostructuring of metals by severe plastic deformation for advanced properties. Nat. Mater. **3**(8), 511–516 (2004)
7. Koch, C.C., Youssef, K.M., Scattergood, R.O., Murty, K.L.: Breakthroughs in optimization of mechanical properties of nanostructured metals and alloys. Adv. Eng. Mater. **7**(9), 787–794 (2005)
8. Estrin, Y., Vinogradov, A.: Extreme grain refinement by severe plastic deformation: a wealth of challenging science. Acta Mater. **61**(3), 782–817 (2013)
9. Valiev, R.Z., Estrin, Y., Horita, Z., Langdon, T.G., Zehetbauer, M.J., Zhu, Y.T.: Fundamentals of superior properties in bulk nanospd materials. Mater. Res. Lett. **3831**, 1–21 (2015)
10. Langdon, T.G.: Twenty-five years of ultrafine-grained materials: achieving exceptional properties through grain refinement. Acta Mater. **61**(19), 7035–7059 (2013)
11. Wang, Y., Chen, M., Zhou, F., Ma, E.: High tensile ductility in a nanostructured metal. Nature **419**(6910), 912–915 (2002)
12. Dhal, A., Panigrahi, S.K., Shunmugam, M.S.: Influence of annealing on stain hardening behaviour and fracture properties of a cryorolled Al 2014 alloy. Mater. Sci. Eng. A 229–238 (2015)
13. Dhal, A., Panigrahi, S.K., Shunmugam, M.S.: Insight into the microstructural evolution during cryo-severe plastic deformation and post-deformation annealing of aluminum and its alloys. J. Alloys Compd. **726**, 229–238 (2017)
14. Srinivas, B., Dhal, A., Panigrahi, S.K.: A mathematical prediction model to establish the role of stacking fault energy on the cryo-deformation behavior of FCC materials at different strain levels. Int. J. Plast. **97**, 159–177 (2017)

15. Thangaraju, S., Heilmaier, M., Murty, B.S., Vadlamani, S.S.: On the estimation of true Hall-Petch constants and their role on the superposition law exponent in al alloys. Adv. Eng. Mater. **14**(10), 892–897 (2012)
16. Kals, T., Eckstein, R.: Miniaturization in sheet metal working. J. Mater. Process. Technol. **103**, 95–101 (2000)
17. Vollertsen, F., Hu, Z., Niehoff, H.S., Theiler, C.: State of the art in micro forming and investigations into micro deep drawing. J. Mater. Process. Technol. **151**, 70–79 (2004)
18. Geiger, M., Klinger, M., Eckstein, R., Tiesler, N., Engel, U.: Microforming. CIRP Ann. Manuf. Technol. **2**, 445–462 (2001)
19. Hug, E., Keller, C.: Intrinsic effects due to the reduction of thickness on the mechanical behaviour of nickel polycrystals. Metall. Mater. Trans. A **41**, 2498–2506 (2010)
20. Deng, J.H., Fu, M.W., Chan, W.L.: Size effect on material surface deformation behavior in micro-forming process. Mat. Sci. Eng. **528**, 4799–4806 (2011)
21. Chan, W.L., Fu, M.W., Yang, B.: Experimental studies of the size effect affected microscale plastic deformation in micro upsetting process. Mater. Sci. Eng. A **534**, 374–383 (2012)
22. Dai, Y.Z., Chiang, F.P.: On the mechanism of plastic-deformation induced surface-roughness. J. Eng. Mater. Technol. Trans. ASME **114**, 432–438 (1992)
23. Dhal, A., Panigrahi, S.K., Shunmugam, M.S.: Development and characterisation of fine-grained aluminium for micro sheet metal forming operation. In: Proceedings of 6th International and 27th All India Manufacturing Technology, Design and Research Conference, Pune, pp. 1497–1500 (2016). ISBN: 978-93-86256-27-0

Chapter 3
Electromagnetic Pulse Crimping of Al-Tube on DP Steel Rod

Ramesh Kumar and Sachin D. Kore

Abstract The electromagnetic pulse crimping is a high energy, high strain rate, high velocity, and green materials joining or surface coating technique. Joining of dissimilar materials is difficult due to their physiochemical properties that are seldom compatible or similar. Therefore, electromagnetic pulse crimping which is solid-state joining technique can be an alternative for joining dissimilar materials. In the present work, composite rods were produced by the electromagnetic pulse crimping technique, which was characterized by a uniform crimping of the flyer tube on the base rod perimeter. The materials used were Al 1050 as flyer tube and dual-phase (DP) steel as a base rod. Numerical simulations were carried out for finding out the optimized parameters for crimping and then experiments were conducted on the optimized parameters. The results obtained from the simulations revealed that for the successful crimping, a minimum value of collision velocity, plastic strain, electromagnetic pressure, and standoff distance must be maintained. The post-process current obtained from the simulations and first peak of the discharge current measured in the experiments was compared. The variation in the maximum value of discharge currents in simulations from the experimental values was found to be 2, 3, and 7% at 2.5, 2.6, and 2.9 kJ of discharge energy. The outer diameter of the successfully crimped samples was measured and compared with the outer diameter obtained from the simulations and found a maximum of 6.6% variation in the simulation value from the experimental value. The optical microscope image was analyzed and it was found that the Al-tube was crimped on the DP steel rod with a negligible gap. Further, pullout tests and hardness tests at the interface were performed to test the strength and hardness of the joints, respectively.

Keywords Electromagnetic pulse crimping · High-velocity impact
Pullout test · Finite element method · Joining by forming

R. Kumar (✉) · S. D. Kore
Indian Institute of Technology Guwahati, Guwahati 781039, Assam, India
e-mail: rems2012mt0088@gmail.com

S. D. Kore
e-mail: sdk@iitg.ernet.in

© Springer Nature Singapore Pte Ltd. 2019
U. S. Dixit and R. G. Narayanan (eds.), *Strengthening and Joining by Plastic Deformation*, Lecture Notes on Multidisciplinary Industrial Engineering,
https://doi.org/10.1007/978-981-13-0378-4_3

3.1   Introduction

The electromagnetic principles were used and demonstrated in electromagnetic pulse joining process. This process fulfills the needs involving dissimilar materials joining and also joining of difficult-to-weld material [1]. Electromagnetic pulse crimping is a high speed, impact joining process which has potential to significantly reduce weight and manufacturing cost, especially for joining cylindrical structures of similar or dissimilar materials. Electromagnetic pulse joining uses an electromagnetic force to accelerate flyer tube and collide against the base rod or tube with very high speed, and produces a solid-state joint without any thermal distortions and external heat source [2].

Electromagnetic pulse joining is not only the useful joining processes for the dissimilar metallic materials joining for cylindrical parts but also a new technology for metallic materials joining by means of the repulsive force between the primary current in the working coil and current induced in an outer tube [3]. The factors affecting the quality of joining are the discharge energy, the gap between the flyer tube and inner tube or rod and thickness of an outer tube or flyer tube. The electromagnetic pulse joining process greatly minimizes the metallurgical incompatibilities problems between dissimilar metals and alloys. Electromagnetic pulse crimping process was developed for components with distinct geometries to utilize the various extents of high pressure to create reliable joints [4].

Electromagnetic pulse crimping is a high energy, high strain rate joining by forming technique [5]. It was used for joining either homogeneous or heterogeneous material having different or same thicknesses [6]. It is a solid-state electromagnetic joining process in which cylindrical components, such as tubes, rods, and cones were joined together in lap configuration using electromagnetic force [7]. In this method, no other pressurizing machine is required and it is an easier process and has the capability of joining within a few microseconds. Therefore, electromagnetic pulse crimping process can be used as a technique which can be used for on-site tube work operations. Electromagnetic pulse crimping process has good repeatability and depending upon impact velocity or discharge energy, the strength of the joint varies [8].

The manufacturing of modern lightweight structures requires appropriate joining technologies for application of multi-material concepts. To fulfill the demand of weight reduction in different engineering structures, the process should have the capacity to join dissimilar metals without additional chemical binders, mechanical elements, or adverse influence of heat on the two or more joining partners. The electromagnetic pulse joining is an alternative to available conventional mechanical or thermal joining process [9]. The electromagnetic pulse joining is normally applicable to lap joint of tube-to-tube or tube-to-bar configuration using the magnetic repulsion between two opposing magnetic fields to drive a conductive metal. The magnetic field produced by the coil penetrates the workpiece, i.e., flyer tube and generates an eddy current in the flyer tube, i.e., workpiece which produces a magnetic field having opposing nature. These two magnetic fields at the coil and the workpiece induce a repulsive force between the working coil and the flyer tube [10].

In electromagnetic pulse joining, energy is stored and discharged in a very short period of time. For example, some electromagnetic systems can discharge in a range of kiloamperes in less than 100 μs of time. The actual expenditure for energy is low because a high amount of electromagnetic energy can be discharged in very short time [11]. In this process, the discharge current passes through a coil which surrounds the workpiece to be joint. The working coil, which wraps around the components, does not come into contact with them. The high discharge current in the working coil produces an eddy current in the outer flyer tube. In the assembly, the coil is fixed and the flyer tube is free to move. Hence, the flyer tube moves away from the coil with a high-velocity impact on the base rod. The manufacturing rate of the joint by this process can be limited between two and ten seconds. For production of joints by this process, fillers, gases, fluxes, and other external materials are not used and the quality of the joint is independent.

The electromagnetic processing system consists of an AC power supply unit, switches, capacitor bank, and coil tool. Different components associated with the electromagnetic pulse crimping unit are shown in Fig. 3.1. In Fig. 3.1, C_1, C_2, and C_3 are the capacitances of the three capacitors, R_m and L_m are the resistance and inductance and of the machine, and L_c and R_c are the inductance and resistance of the coil. Also, R_w and L_w are the resistance and inductance of the workpiece. Coil tool is the actual tool, with the help of which magnetic pressure is being generated on the flyer tube. In Fig. 3.1, M represents the mutual inductance of the current from the coil to the workpiece. The currents flowing in the primary and secondary circuit are coil current ($I_c(t)$) and workpiece current ($I_w(t)$). There are two switches, switch A and switch B in the circuit, shown in Fig. 3.1.

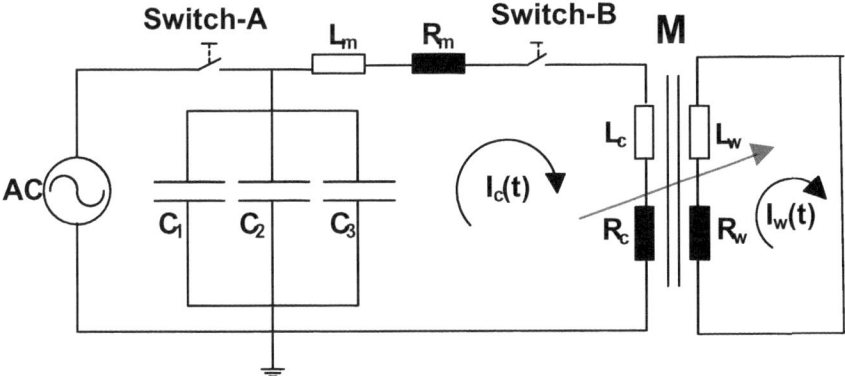

Fig. 3.1 The circuit diagram of the electromagnetic pulse crimping process

magnetic force $\overrightarrow{F}$ in unit volume of surrounding medium, i.e., the magnetic force density, can be calculated by Eq. (3.9).

$$\overrightarrow{F} = \overrightarrow{J} \times \overrightarrow{B} = \frac{1}{\mu}\left(\text{curl } \overrightarrow{B}\right) \times \overrightarrow{B} \tag{3.9}$$

where $\overrightarrow{J}$ is the current density (in A/m^2) and $\overrightarrow{B}$ is the magnetic flux density (T).

3.2.2 Finite Element Method

Finite element method was used for this analysis which deals with high strain rate forming. For this high strain rate forming strain rate sensitivity, Johnson–Cook plasticity model was used. In the used simplified Johnson–Cook plasticity model, thermal effects and thermal damages are ignored. Since thermal softening is not considered in this model which is very significant in reducing the yield stress with adiabatic loading, there is limited maximum stress. This model is 50% faster than the Johnson–Cook materials model because of the ignorance of the thermal damage and thermal softening. To compensate the deficiency of thermal softening, the stress values used were limited and the stresses were within reasonable limits.

To model the behavior of the tubes in the finite element simulation, the Johnson–Cook constitutive equation was used. The Johnson–Cook equation which is a combination of plastic strain rate and plastic strain can be described by Eq. (3.10).

$$\sigma = (A + B\varepsilon^n)(1 + C \ln \dot{\varepsilon})\left(1 - \left(\frac{T - T_r}{T_m - T_r}\right)^m\right) \tag{3.10}$$

where σ represents the flow stress of the material, ε represents an equivalent plastic strain, $\dot{\varepsilon}_\rho$ represents plastic strain rate, T represents the absolute temperature, T_r represents the room temperature, T_m represents the melting temperature of the material, and A, B, C, n, m are Johnson–Cook constants. Constants in this equation for materials Al-tube and DP steel rod are given in Table 3.1. For EM simulations, the model was simulated for different values of discharge voltages, ranging from 6.8, 7.0, 7.1, 7.4, 7.7, and 8.0 kV.

Table 3.1 Values of Johnson–Cook material constant parameters

Materials	A (MPa)	B (MPa)	n	C	T_m (K)	m
Al 1050 [17]	110	150	0.4	0.01	918	1
DP steel [18]	430	823.6	0.5	0.08	1048	0.6

3.3 Experimental Detail

The electromagnetic pulse crimping was carried out using a maximum of 10 kJ rated energy and 15 kV rated voltage electromagnetic pulse processing system. In this section, the types of materials used, the initial input conditions, and the input load which was the discharge current were discussed in detail.

3.3.1 Materials and Method

Electromagnetic pulse crimping of Al 1050 tube (tube dimensions: 17 mm outer diameter, 0.6 mm thickness, and 40 mm length) to dual-phase (DP) steel rod (rod dimensions: 15 mm outer diameter and 40 mm length) was performed using an electromagnetic processing system with a maximum charging energy of 10 kJ and voltage of 15 kV. The total capacitance of the capacitors equals 90 μF. Figure 3.3 shows the overlap configuration and corresponding dimensions of the Al-tube and DP steel rod in the helical coil. The Al-tube is called the flyer tube and is located just below the coil conductor. The DP steel rod is called base rod. The distance by which the Al flyer tube is separated from the DP steel base rod prior to the discharge is the standoff distance. The overlap is the distance between the spiral coil and the Al flyer tube. The mechanical properties of the flyer tube and base rod are listed in Table 3.2 and their chemical compositions are listed in Table 3.3.

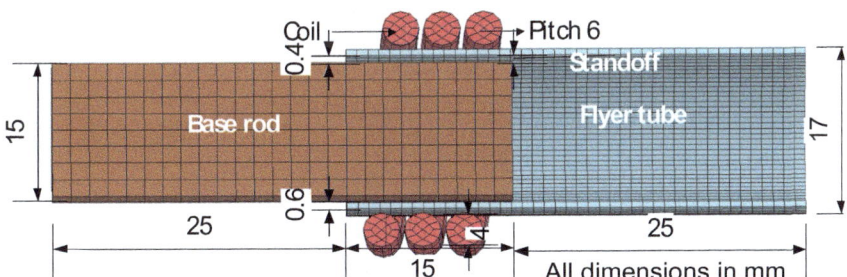

Fig. 3.3 Dimensions of the copper coil, Al-tube, and DP steel rod sample

Table 3.2 Mechanical properties of the materials used in the study

Materials	Ultimate tensile strength (MPa)	Yield tensile strength (MPa)	Elongation at break (%)	Density (g/cc)
DP steel	500–600	300–380	≥ 25	7.870
Al 1050	160	145	≈7.0	2.705

Table 3.3 Chemical composition of the materials used

Materials	C (%)	Fe (%)	Mn (%)	Si (%)	Al (%)	Ti (%)	V (%)
DP steel	≤ 0.14	≥ 97.86	≤ 1.6	≤ 0.40	–	–	–
Al 1050	–	≤ 0.40	≤ 0.1	≤ 0.25	≥ 99.5	≤ 0.1	≤ 0.1

3.3.2 Process Conditions

In the simulations, the discharge energy was varied from 1.0 to 2.9 kJ. The standoff distance was fixed throughout the study, and it was equal to 0.4 mm about half of the thickness and overlap distance was taken as 15 mm.

Generally, three types of the coil were used in the electromagnetic pulse compression joining for cylindrical configuration, namely single-turn coil, multi-turn solenoid coil, and multi-turn plate coil. In the single-turn coil, there is only one turn or winding around the workpiece. But in the multi-turn coil, the number of turns or winding around the sample is either two or more than two. The multi-turn solenoid coil is made of a single circular wire in solenoid shape but in the multi-turn plate coil, the coil is made of a number of plates made of either aluminum or copper. In this study, a three-turn solenoid coil was used for compression joining.

The lap configuration joining of the samples can be achieved in three ways, such as middle joint, left end joint, and right end joint. In middle joint configuration, the tube was compressed at other end of the tube but in the end joint, the tube was compressed at the end position. In this study, the lap configuration obtained was left end joint. The input load applied to the electromagnetic pulse crimping is the peak value of the discharge current. Mainly, the first peak of the discharge current was responsible for the deformation of the flyer tube which causes joining with the base rod. Discharge current at different energies was measured using Rogowski coil and oscilloscope. Figure 3.4 shows the discharged current curve and Fig. 3.5 shows the first peak of the discharged current measured during the experiment at different discharge energies. The mathematical relation between the discharge current curve can be expressed by Eq. (3.11) [5].

$$I(t) = \frac{V_o}{\omega L} e^{-\beta t} \sin(\omega t) \tag{3.11}$$

In the electromagnetism model used, damped sinusoidal current was used as a load which passes through the coil tool and it can be expressed in the mathematical form by Eq. (3.12) [5]:

$$I = V_o \sqrt{\left[\frac{C}{L} e^{-\beta t} \sin(\omega t) \right]} \tag{3.12}$$

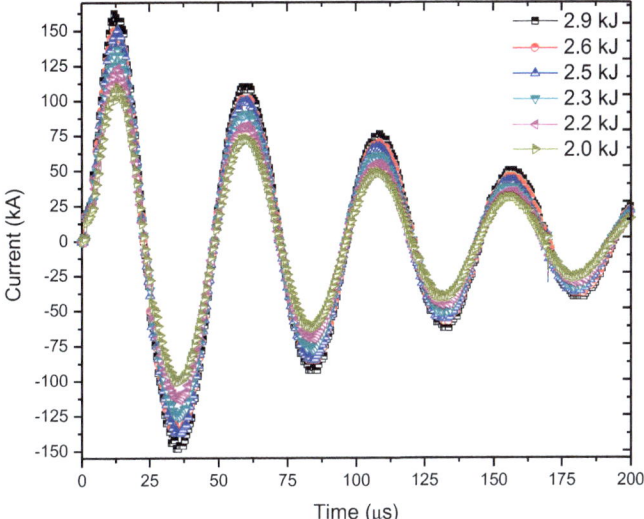

Fig. 3.4 Discharge current curve obtained from the experiment

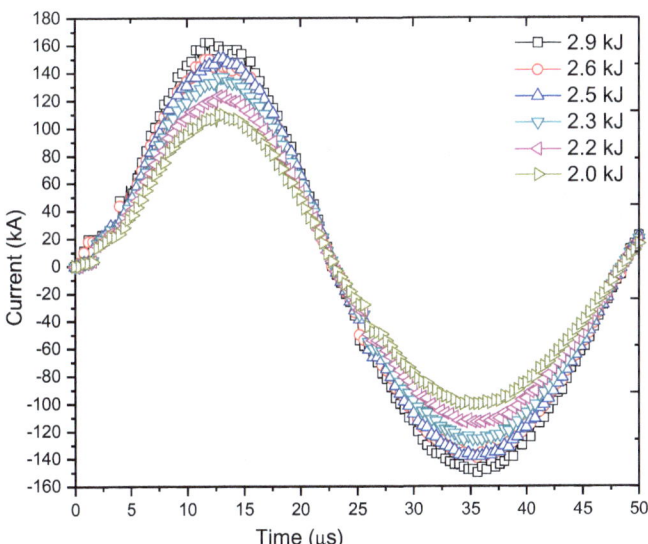

Fig. 3.5 Enlarged view of the first peak of the measured discharge current

where V_o represents the initial discharged voltage, represents the damping factor or exponent, L represents the equivalent inductance of the circuit in the system [8], C represents the total capacitance of the capacitor bank, and ω represents the angular frequency.

The damping coefficient can be calculated using Eq. (3.13) and the angular frequency of the current can be calculated by Eq. (3.14).

$$\beta = \frac{R}{2L} \tag{3.13}$$

$$\omega = \sqrt{\left(\omega_0^2 - \beta^2\right)} \tag{3.14}$$

where ω_0 is the initial angular frequency and it can be calculated by Eq. (3.15) and R, L, and C_c are the resistance, inductance, and capacitance of the circuit, respectively [5].

$$\omega_0 = \frac{1}{\sqrt{LC_c}} \tag{3.15}$$

3.3.3 Pullout Test

The samples were crimped successfully at six different discharge energies of 2.00, 2.20, 2.30, 2.50, 2.60, and 2.90 kJ with three-turn solenoid coil. To evaluate the joint strength in tension, the pullout test of the aluminum tube crimped on the DP steel rod was subjected to tensile load, i.e., pulling load. In a total of 12 samples, two samples at each discharge energy were tested. Finally, the joint strength was evaluated by performing the tests using the universal testing machine (UTM) at a speed of 0.5 mm/min. The schematic diagram of the pullout test performed on the UTM is shown in Fig. 3.6. In this test, crimped length was 15 mm and gripping length was 18 mm in the UTM gripper.

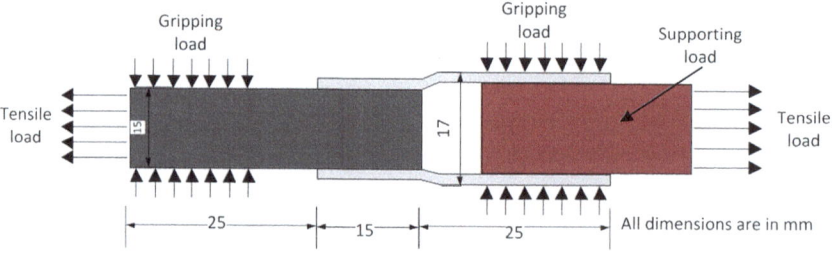

Fig. 3.6 Pullout test arrangement made

3.4 Results and Discussions

The numerical simulations, as well as the experiments, were performed to obtain the successful crimping of the Al-tube on the DP steel. The input parameters which greatly affect the process were the radial gap and the discharge energy. The various post-process results obtained from the process are discussed in the following sections.

3.4.1 Simulation Results

Electromagnetic pulse crimping is a complex transient high-velocity impact joining process. It involves the coupling effects of an electromagnetic field, thermal field, and mechanical field. In this work, the electromagnetism module of LS-DYNA was used for coupling the electromagnetic, structural, and mechanical process. In this module, source electrical current can be introduced into the ends of the helical coil to solve the problem related to the electromagnetic field, structural field, and thermal field. Three-dimensional meshed numerical model for electromagnetic crimping is shown in Fig. 3.7. Ten simulations were run from 1.0 to 2.9 kJ and out of which six simulations were found suitable for the study. Four simulations run below 2 kJ were not considered because the deformation obtained in the flyer tube was not found to be sufficient.

3.4.1.1 Deformation and Impact Velocity

Deformation of the flyer Al-tube was simulated at different energies. Changes in the diameter of the flyer tube at six different time steps are shown in Fig. 3.8. The outer diameter of the flyer tube crimped on the base rod at six different discharge energies was measured in the simulations as well as in the experiment. The outer diameter of

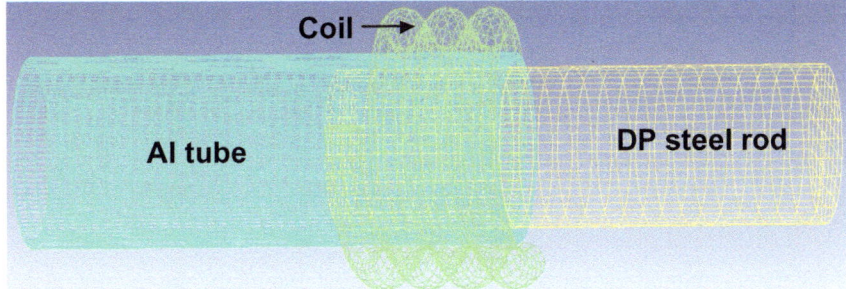

Fig. 3.7 Three-dimensional meshed model used for the coupled field analysis

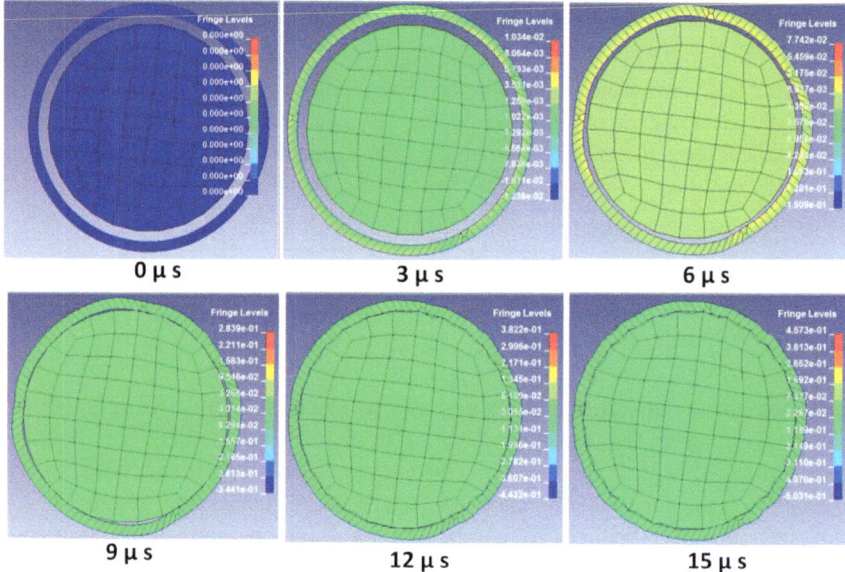

Fig. 3.8 Contours of resultant displacement at six different time steps in the simulations

the crimped samples both in the experiment and in the simulations is compared and shown in Fig. 3.9. The diameter of the rod and tube was also plotted for taking the reference for comparison.

The measured outer diameter of the experiment is shown in Fig. 3.10 and the simulation is shown in Fig. 3.11. There was a variation in the measured diameter of the crimped samples in simulation and in the experiment, it was an acceptable

Fig. 3.9 Comparison of outer diameter from the simulation and experiment

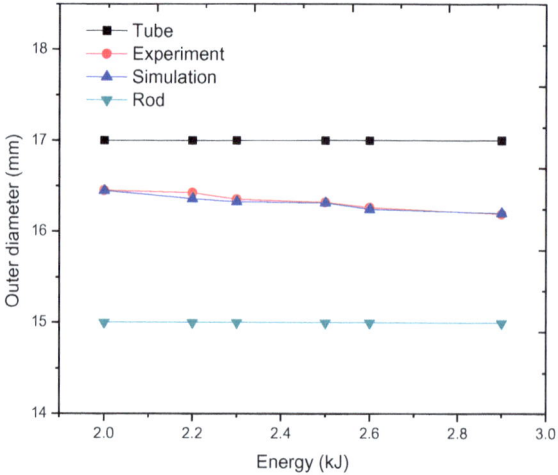

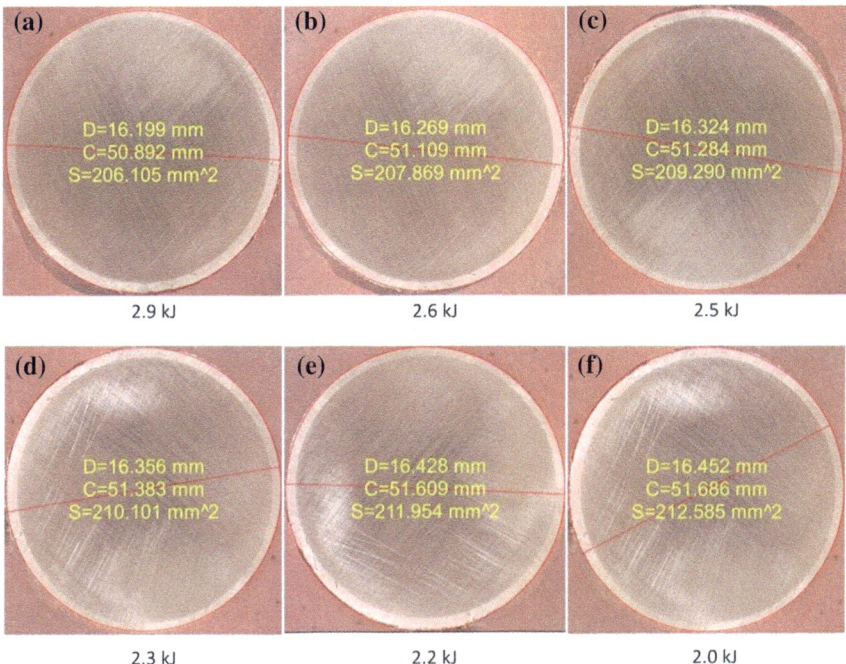

Fig. 3.10 Outer diameter of the crimped sample at six different discharge energies

range. There was a variation in the measured outer diameter of the crimped samples from the experiment and in the simulation. A maximum of 6.6% variation was observed in the simulation data with respect to the experiment value. The percentage variation in the outer diameter at six different values of the discharge energies is tabulated in Table 3.4.

With the increase in the discharge energy, the deformation or displacement of the flyer tube also increases. Figure 3.12 shows the variation of the displacement with respect to time for six different discharge energies. From Fig. 3.12, it can be concluded that with respect to time the deformation of the tube first increases, attained the maximum value, and then decreases. The maximum deformation obtained was corresponding to the peak value of the discharge current.

In electromagnetic crimping, impact velocity is a very important process parameter. In the simulations, the impact velocities under different discharge energies were studied. The average value of simulated impact velocities under six different discharge energies is shown in Fig. 3.13. It can be concluded from Fig. 3.13 that with the increase in the discharge energy the magnitude of the impact velocity also increases. The mechanical strength of the joint was greatly influenced by the amount of the impact velocity; the greater the impact velocity, the higher will be the strength of the joint. It can also be concluded from Fig. 3.13 that for a particular value of the discharge energy, the magnitude of the impact velocity first

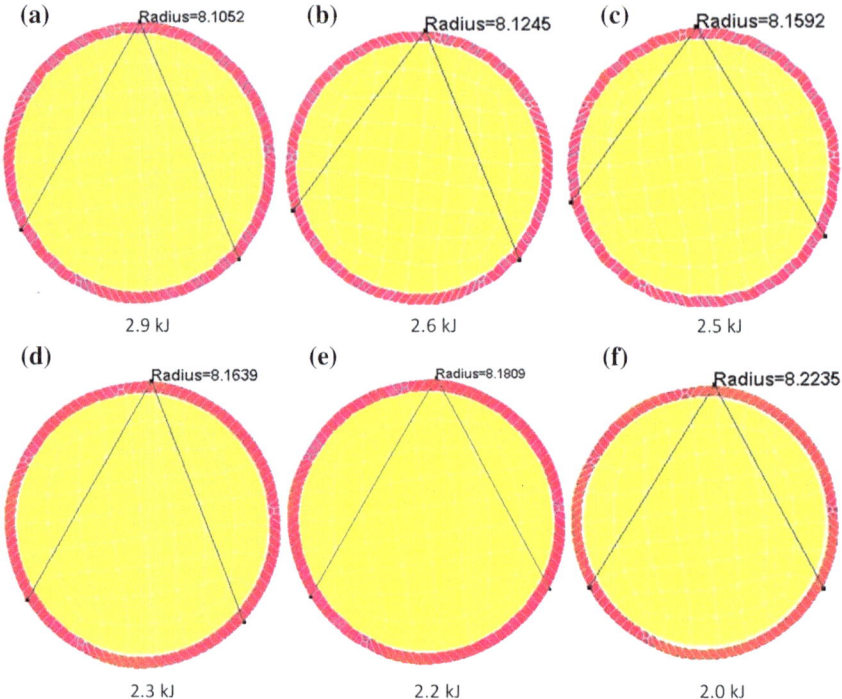

Fig. 3.11 Outer diameter of the crimped sample at six different discharge energies

Table 3.4 Comparison of outer diameter in the experiment and simulation

Energy	2.00 kJ	2.20 kJ	2.30 kJ	2.50 kJ	2.60 kJ	2.90 kJ
Outer diameter from experiment	16.452	16.428	16.356	16.324	16.269	16.199
Outer diameter from simulation	16.447	16.362	16.328	16.318	16.249	16.210
% variation from experiment	0.5	6.6	2.8	0.6	2	−1.1

increases with the time and then decreases. The maximum value of the impact velocity was found corresponding to the peak value of the discharge current.

3.4.1.2   Current and Magnetic Pressure

The discharge current and the magnetic pressure were also analyzed in the study. The maximum value of the discharge current obtained from the simulation and that from the experiment have variation but in an acceptable range. The discharge current

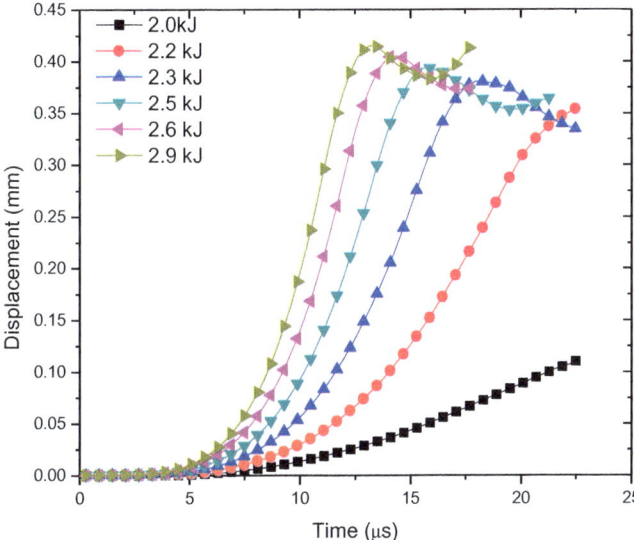

Fig. 3.12 Simulated displacements under different discharge energies

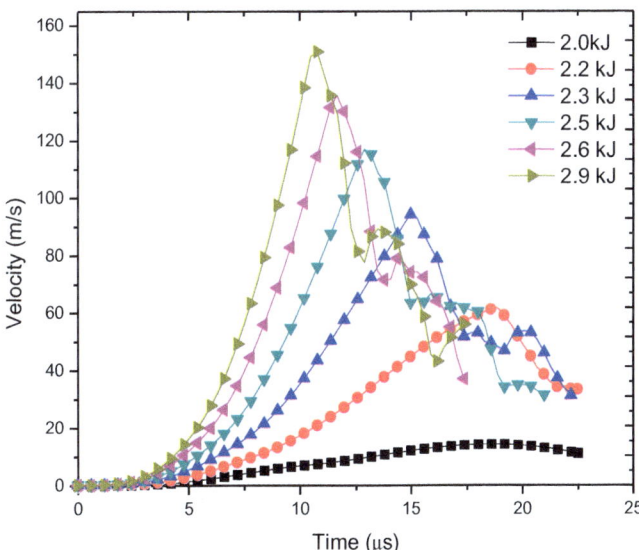

Fig. 3.13 Simulated velocities under different discharge energies

obtained from the simulations is shown in Fig. 3.14. The variation in the maximum value of discharge currents in simulations from the experimental values was found to be 2, 3, and 7% at 2.5, 2.6, and 2.9 kJ of discharge energy, respectively.

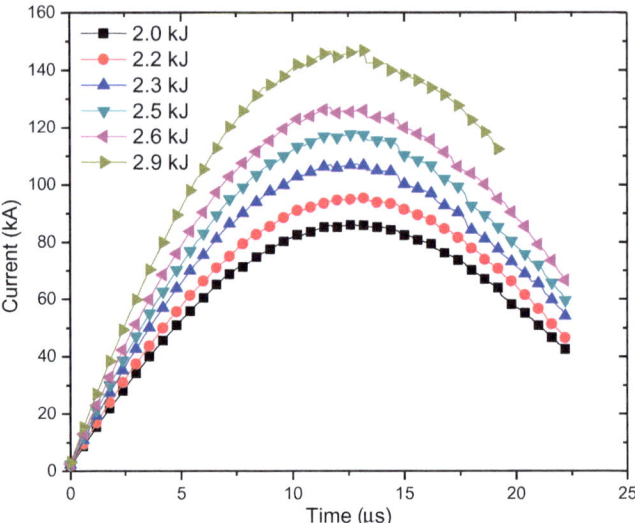

Fig. 3.14 Simulated currents under different discharge energies

The maximum current produced in this process by a capacitor bank is directly proportional to the charging voltage (V_o) and the square root of the capacitance (C), and also inversely related to the square root of the inductance (L) [19]. Mathematically, the peak current produced can be expressed by Eq. (3.16) and the current density J is given by Eq. (3.17).

$$I_{max} = V_o \sqrt{\frac{C}{L}} \tag{3.16}$$

$$J = n\frac{I}{A} \tag{3.17}$$

where A represents the cross-sectional area of the wire of the actuator and n represents the number of turns of the actuator. From the simulation, the contour plot of the current density generated on the flyer tube at six different time steps at 2.90 kJ of discharge energy is shown in Fig. 3.15. In Fig. 3.15, the unit of the current density is mA/mm^2. The maximum magnitude of the current density was observed at about 14 µs, and then it was decreasing with the increase in the time.

The magnetic pressure variation with the time in the simulation at six different discharge energies is shown in Fig. 3.16. It was found that with the increase in the amount of the discharge energy, the magnitude of the magnetic pressure also increases. The increase in the magnetic pressure causes the increase in the impact velocity of the flyer tube. It was also observed that first with the increase in time the

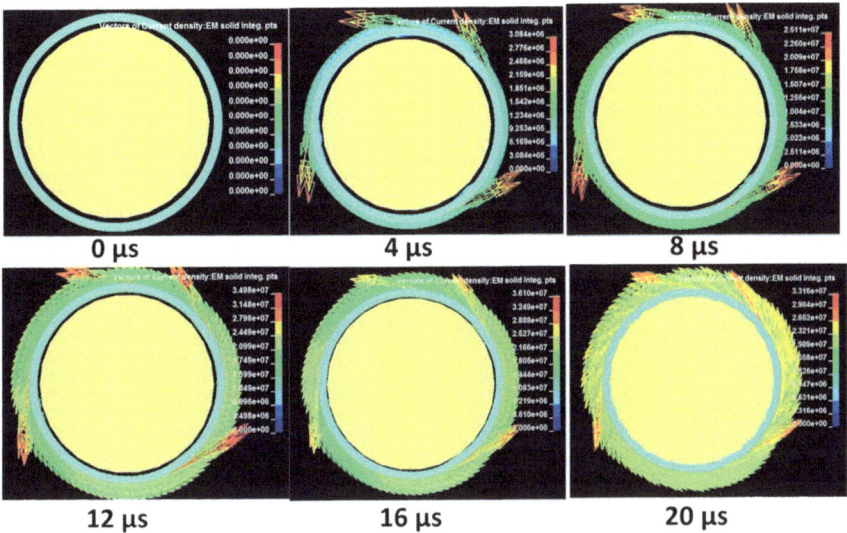

Fig. 3.15 Contour plot of the current density generated on the flyer tube at different time steps

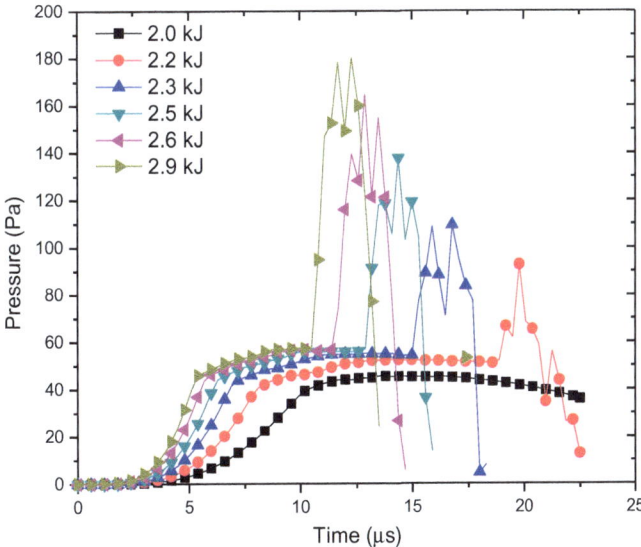

Fig. 3.16 Simulated magnetic pressures under different discharge energies

magnitude of the pressure increases, attains a maximum amplitude, and then decreases with the time. The amplitude of the magnetic pressure was maximum corresponding to the maximum amount of applied load, i.e., maximum discharge current.

3.4.1.3 Tresca Shear Stress and Magnetic Field

The variation of the Tresca maximum shear stress developed in the flyer tube in the simulations was studied. It was found that with the increase in the discharge energy the shear stress value also increases. The variation in the maximum shear stress in the tube with the simulation time at six different discharge energies is shown in Fig. 3.17. The maximum value of the Tresca shear stress developed at six different discharge energies is listed in Table 3.5. The magnetic force generated on the flyer tube is a proportion of the magnetic field strength. The distribution of the magnetic field of the flyer tube was studied in the simulation. Figure 3.18 shows the simulated results of the magnetic field under six different discharge energies.

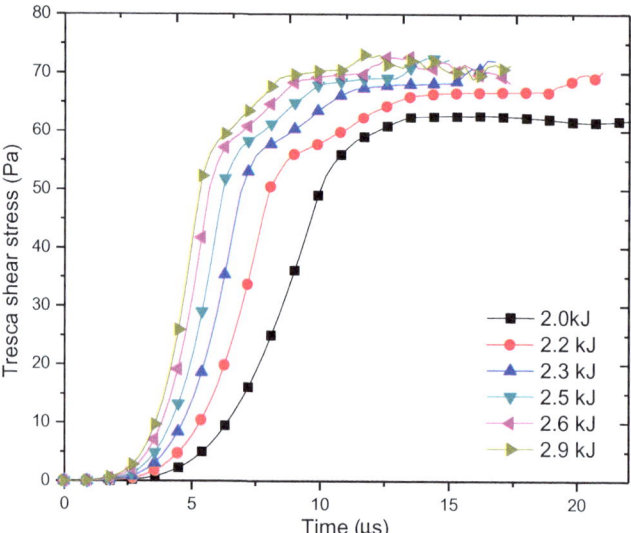

Fig. 3.17 Tresca shear stress in tube at different energies

Table 3.5 Simulation results at different process parameters

Energy	2.00 kJ	2.20 kJ	2.30 kJ	2.50 kJ	2.60 kJ	2.90 kJ
Max. current (kA)	85.82	95.31	106.7	117.898	126.12	146.8
Max. disp. (mm)	0.112	0.357	0.38	0.393	0.405	0.415
Max. plastic strain	0.009	0.053	0.099	0.140	0.173	0.220
Max. pressure (MPa)	45.47	93.22	109.77	137.92	164.98	180.24
Max. lorentz force (N)	16.95	34.47	58.71	89.49	125.09	167.68
Max. magnetic field (Tesla)	5.06	7	9.18	11.4	13.62	15.86
Max. velocity (m/s)	14.36	61.55	94.18	116.83	135.6	150.89
Tresca max. shear stress (MPa)	62.6	70	72	72.5	72.8	73.1

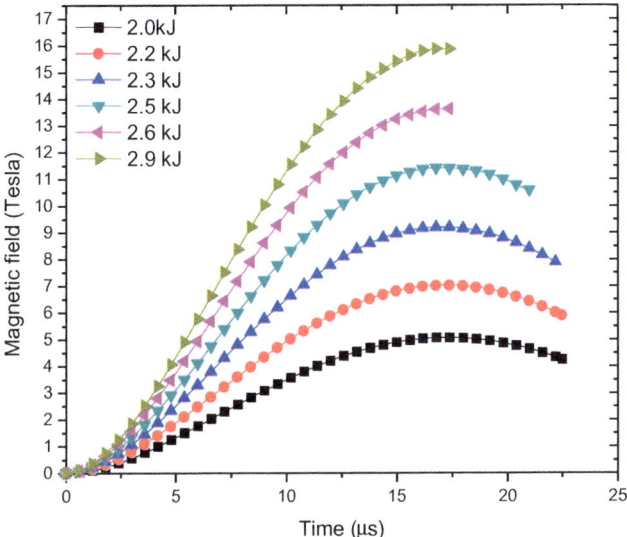

Fig. 3.18 Magnetic field under six different discharge energies

The maximum values of the deformation, impact velocity, discharge current, magnetic pressure, magnetic field strength, and plastic strain at six different discharge energies are listed in Table 3.5.

3.4.1.4   Plastic Strain

Plastic strain in the flyer tube as well as in the rod was also studied in the simulations. In literature, it was stated that for getting successful crimping or joining, minimum value of plastic strain must be attained [20]. In the simulations, the maximum value of plastic strain obtained was 0.23 for 2.9 kJ discharge energy. The contours of the plastic strain in the flyer and in the rod are shown in Figs. 3.19 and 3.20, respectively. Figure 3.21 shows the average variation in the plastic strain with the time of the flyer tube, in the region where deformation occurred. The maximum value of the plastic strain at six different discharge energies is listed in Table 3.5. Table 3.5 shows the process parameters, such as the maximum values of the current (kA), displacement (mm), plastic strain, pressure (MPa), Lorentz force (N), magnetic field (Tesla), impact velocity (m/s), and Tresca Max. Shear stress (MPa) is listed at six different energies. It was found that with the increase in discharge energy, the process parameters values also increase.

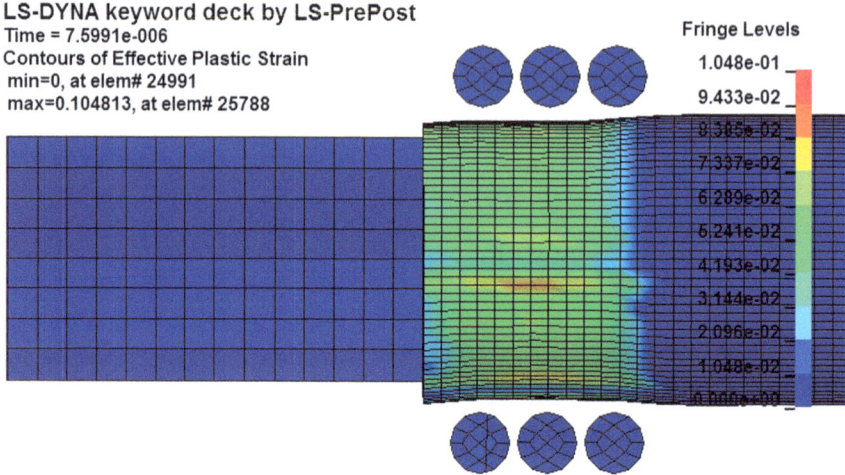

Fig. 3.19 Contours of plastic strain in the tube

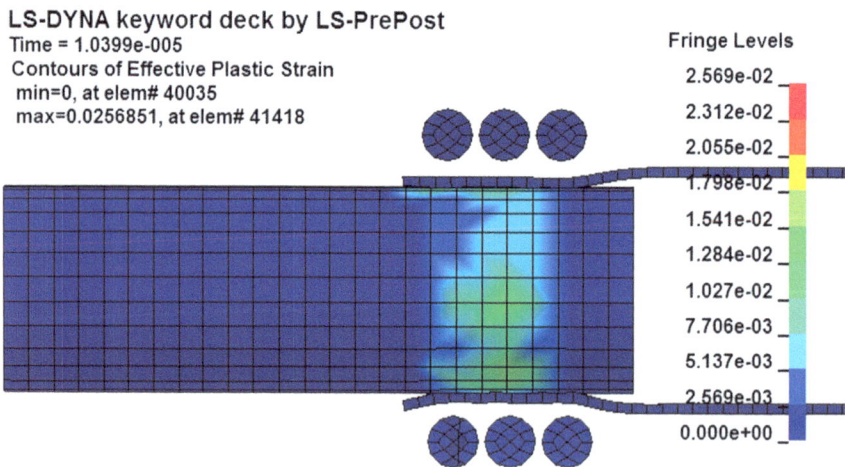

Fig. 3.20 Contours of plastic strain in the rod

3.4.2 Experimental Results

Based on the parameters in the simulations, experiments were performed, and Al-tube was successfully crimped on the DP steel bar. Figure 3.22 shows the samples being crimped successfully. The optical microscope (OM) images at the interface of the sample crimped at 2.9 kJ of discharge energy were analyzed. In the OM image, a negligible gap was found for the sample being crimped at 2.9 kJ

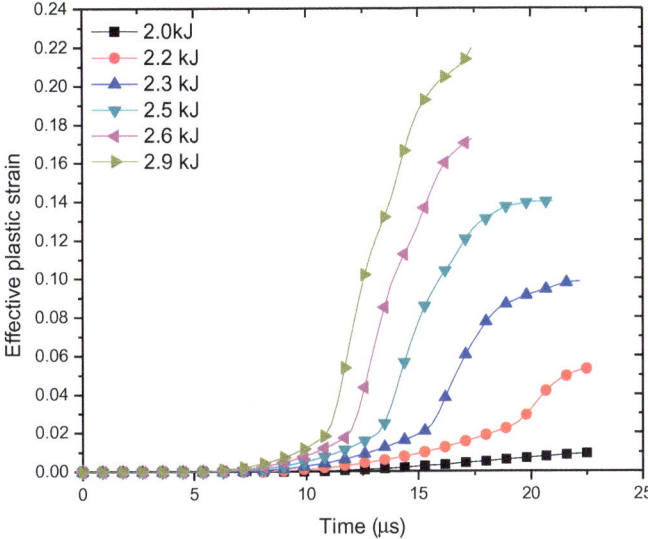

Fig. 3.21 Simulated plastic strains under six different discharge energies

Fig. 3.22 Al-tube crimped over DP steel rod

discharge energy. With the increase in the discharge energy, it was found that the quality of crimping was improved.

Different types of the pattern were observed at different locations in the OM images of the electromagnetic pulse crimped samples. Some of the specific interface

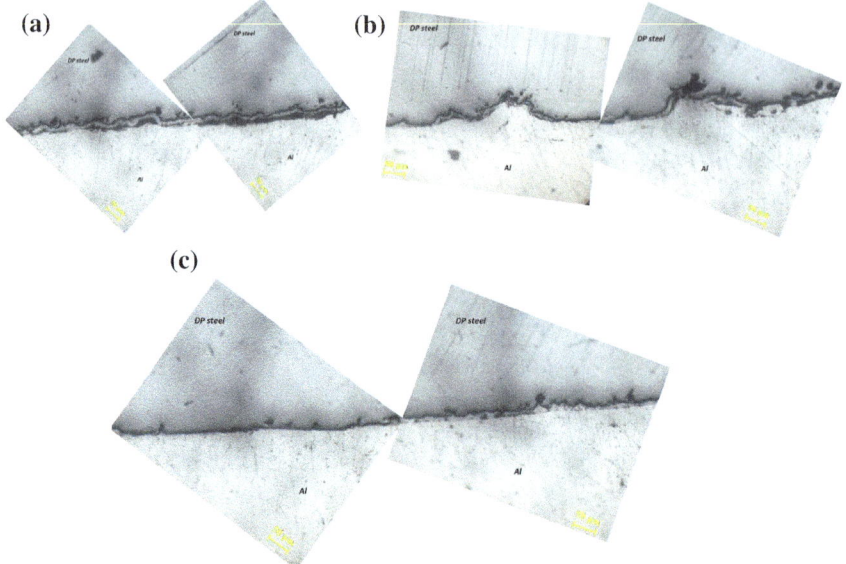

Fig. 3.23 Small wavy interface morphology along crimping interface of the crimped sample, **a** small wavy, **b** large wavy, and **c** straight pattern

morphologies, such as small wavy, large wavy, and straight pattern are shown in Fig. 3.23a–c. These types of morphology at the interface are due to the irregularities available on the DP steel rod. This will result in such types of morphology after the high-velocity impact of the flyer tube on the rod. The strength of the crimped joint was varied with the types of profile or surface roughness available on the surface of the base rod. The effect of the mandrel's surface on the mechanical properties of joints was studied by Hammers et al. and found that surface roughness increases the strength of the joint [21]. Different types of surface profiles also lead to varying the mechanical joint strength of the crimped sample. The influence of three types of surface profiles, namely plain, knurled, and threaded on the base was studied by Kumar and Kore and found that threaded profile results in better strength in comparison to other two profiles [22].

3.4.3 Hardness Test

Hardness tests using Vickers' microhardness tester with loads 300 and 500 gf were performed. Hardness was measured at four different locations on the sample crimped at 8 kV or 2.9 kJ discharge energy. Figure 3.24 shows an average hardness traverse across an Al/DP steel clad interface. The increase in hardness near the interface base materials is due to the high-velocity impact.

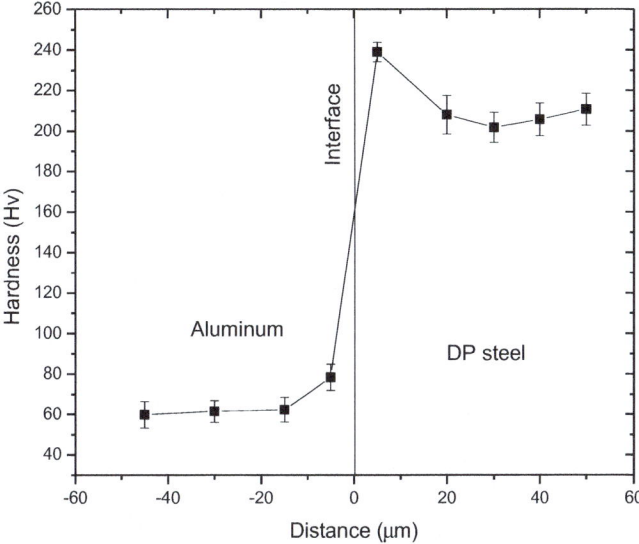

Fig. 3.24 Hardness traverse across Al and DP steel crimped interface

3.4.4   Pullout Test

The variation in the pullout load and corresponding extension obtained in the joined sample at six different discharge energies is shown in Fig. 3.25. Pullout test results revealed that all the samples joined up to 2.9 kJ of discharge energy fail with

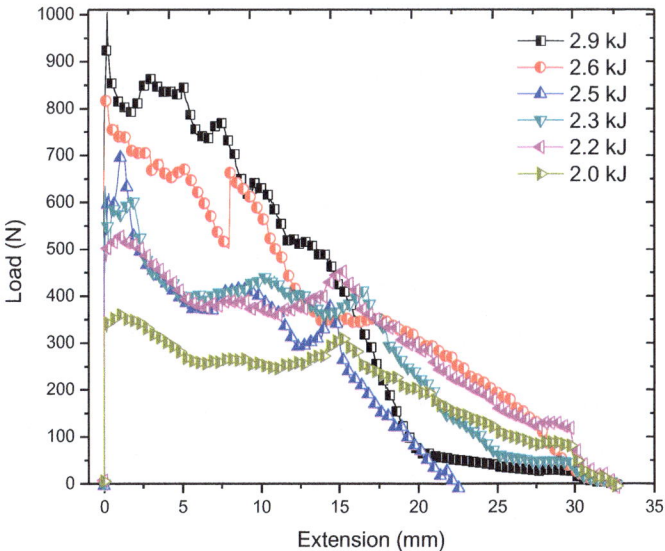

Fig. 3.25 Pullout load versus extension plot at six different discharge energies

separation failure mode. In the separation failure mode, the two samples joined together were separated from each other from the joint. The maximum pullout load required was increased with the increase in discharge energy used to join the sample. With the increase in the discharge energy, the strength of the joint was increased, due to the increase in the impact velocity of the flyer tube.

3.5 Conclusions

Electromagnetic pulse joining is a high-velocity forming technique. In this work, electromagnetic pulse crimping was simulated for ten different discharge energies out of which six simulations were chosen for the experiments. Experiments were performed based on the chosen simulations parameters and aluminum tube was crimped successfully on the DP steel rod. Simulations results were validated with the experimental results with measured discharge current and the outer diameter of the flyer tube in the crimped sample. A variation in the maximum value of discharge currents in simulations from the experimental values was 2, 3, and 7% at 2.5, 2.6, and 2.9 kJ of discharge energy, respectively. The outer diameter of the successfully crimped samples was measured and compared with the outer diameter of the model in the simulations and found a maximum of 6.6% variation in the simulation value from the experimental value. From the experimental work and simulation results, it can be concluded that process parameters, such as standoff distance, impact velocity, and discharge voltage or discharge energy must have a threshold value of 0.4 mm, 116.83 m/s, and 2.5 kJ, respectively, for achieving successful crimping. Optical microscopy analysis revealed that there was a negligible gap in the sample crimped at 2.9 kJ of discharge energy. Hardness test showed that hardness was increased near the interface base materials due to the high-velocity impact. The strength of the joints produced by crimping Al-tube on the DP steel rods has increased with the increase in discharge energy, and a maximum load obtained in the test was 1015 N for sample crimped at 2.9 kJ.

References

1. Desai, S.V., Kumar, S., Satyamurthy, P., Chakravartty, J.K., Chakravarthy, D.P.: Scaling relationships for input energy in electromagnetic welding of similar and dissimilar metals. J. Electromag. Anal. Appl. **2**, 563–570 (2010)
2. Eide, H.O., Melby, E.A.: Blast loaded aluminium plates experiments and numerical simulations. Master thesis Norwegian University of Science and Technology (2013), pp. 1–139
3. Haiping, Y., Chunfeng, L.: Effects of current frequency on electromagnetic tube compression. J. Mater. Process. Technol. **209**, 1053–1059 (2009)
4. Haiping, Y., Zhisong, F., Chunfeng, L.: Magnetic pulse cladding of aluminum alloy on mild steel tube. J. Mater. Process. Technol. **214**, 141–150 (2014)

5. Hammers, T., Marre, M., Rautenberg, J., Barreiro, P., Schulze, V., Biermann, D., et al.: Influence of Mandrel's surface and material on the mechanical properties of joints produced by electromagnetic compression. Steel Res. Int. **80**(5), 366–375 (2009)
6. Hokari, H., Sato, T., Kawauchi, K., Muto, A.: Magnetic impulse welding of aluminium tube and copper tube with various core materials. Weld. Int. **12**(8), 619–626 (1998)
7. Kimchi, M., Shao, H., Cheng, W., Krishnaswamy, P.: Magnetic pulse welding aluminum tubes to steel bars. Weld. World **48**, 19–22 (2004)
8. Kore, S.D., Date, P.P., Kulkarni, S.V.: Effect of process parameters on electromagnetic impact welding of aluminum sheets. Int. J. Impact Eng. **34**, 1327–1341 (2007)
9. Kore, S.D., Dhanesh, P., Kulkarni, S.V., Date, P.P.: Numerical modeling of electromagnetic welding. Int. J. Appl. Electromag. Mech. **32**(1), 1–19 (2010)
10. Kore, S.D., Imbert, J., Zhou, Y., Worswick, M.: Magnetic pulse welding. ASM Handb. **06A**, 704–709 (2011)
11. Kumar, R., Kore, S.D.: Effects of surface profiles on the joint formation during magnetic pulse crimping in tube-to-rod configuration. Int. J. Precis. Eng. Manuf. **18**(9), 1181–1188 (2017)
12. Lee, J. G., Park, J. J., Lee, M. K., Rhee, C. K., Kim, T. K., Spirin, A., et al. (2015). End closure joining of ferritic-martensitic and oxide-dispersion strengthened steel cladding tubes by magnetic pulse welding. Metall. Mater. Trans. A 1–8
13. Lueg-Althoff, J., Lorenz, A., Gies, S., Weddeling, C., Goebel, G., Tekkaya, A., et al.: Magnetic pulse welding by electromagnetic compression: determination of the impact velocity. Adv. Mater. Res. **966–967**, 489–499 (2014)
14. Mamalis, A.G., Manolakos, D.E., Kladas, A.G., Koumoutsos, A.K.: Electromagnetic forming and powder processing: trends and developments. Appl. Mech. Rev. (ASME) **57**(4), 299–324 (2004)
15. Marya, M., Rathod, M., Marya, S., Kutsuna, M., Priem, D.: Steel-to-Aluminum joining by control of interface microstructures—laser-roll bonding & magnetic pulse welding. Mater. Sci. Forum **539–543**, 4013–4018 (2007)
16. Miranda, R.M., Tomas, B., Santos, T.G., Fernandes, N.: Magnetic pulse welding on the cutting edge of industrial applications. Soldag. Insp. São Paulo **19**(01), 69–81 (2014)
17. Mousavi, S.A., Burley, S.J., Al-Hassani, S.S.: Simulation of explosive welding using the Williamsburg equation of state to model low detonation velocity explosives. Int. J. Impact Eng. **31**, 719–734 (2005)
18. Psyk, V., Risch, D., Kinsey, B.L., Tekkaya, A.E., Kleiner, M.: Electromagnetic forming—a review. J. Mater. Process. Technol. **211**(5), 787–829 (2011)
19. Shim, J.K., Kang, B.Y., Kim, I.S., Kang, M.J., Park, D.H., Kim, I.J.: A study on distributions of electromagnetic force of the dissimilar metal joining in MPW using a FEM. Adv. Mater. Res. **86**, 214–221 (2010)
20. Shim, J.Y., Kim, I.S., Kang, M.J., Kim, I.J., Lee, K.J., Kang, B.Y.: Joining of aluminum to steel pipe by magnetic pulse welding. Mater. Trans. **52**(5), 999–1002 (2011)
21. Spitz, B.T., Shribman, V.: Magnetic pulse welding for tubular applications: discovering new technology for welding conductive materials. TPJ—Tube Pipe J. 1–3 (2000)
22. Vedantam, K., Bajaj, D., Brar, N.S., Hill, S.: Johnson-Cook strength models for mild and DP 590 steels. Am. Inst. Phys. 775–779 (2006)

Chapter 4
Effect of the Post-weld Heat Treatments on Mechanical and Corrosion Properties of Friction Stir-Welded AA 7075-T6 Aluminium Alloy

S. B. Pankade, P. M. Ambad, R. Wahane and C. L. Gogte

Abstract Age hardenable, high-strength aluminium alloys are used majorly in aerospace, defence, marine and automobile components because of their excellent strength-to-weight ratio and better corrosion resistance. To join such nonferrous alloys, friction stir welding (FSW) is a special technique, which uses the phenomenon of friction and deformation while joining in the solid state. The joint sections of 2XXX, 6XXX and 7XXX series of aluminium alloys are susceptible to microstructural changes during FSW due to their ageing characteristics. These changes further aggravate the problem of mechanical engineering properties, and especially the corrosion resistance of these alloys. In the present work, focus is kept on the effect of post-weld heat treatments (PWHT) such as retrogression and re-ageing (RRA) and stabilization with double ageing (SDA) on mechanical properties, electrical conductivity and exfoliation corrosion resistance of AA 7075 aluminium alloy FSW joints.

Keywords Friction stir welding · AA7075 · PWHT · Exfoliation corrosion RRA · SDA · Electrical conductivity

S. B. Pankade (✉) · P. M. Ambad · R. Wahane · C. L. Gogte
Department of Mechanical Engineering, Marathwada Institute of Technology,
Aurangabad 431010 Maharashtra, India
e-mail: sandeep.patil@mit.asia

P. M. Ambad
e-mail: prashant.ambad@mit.asia

R. Wahane
e-mail: riteshaero@gmail.com

C. L. Gogte
e-mail: shekhar.gogte@mit.asia

© Springer Nature Singapore Pte Ltd. 2019
U. S. Dixit and R. G. Narayanan (eds.), *Strengthening and Joining by Plastic Deformation*, Lecture Notes on Multidisciplinary Industrial Engineering,
https://doi.org/10.1007/978-981-13-0378-4_4

4.1 Introduction

The automobile and aerospace industries have been strongly in favour of utilizing high-strength, age hardenable aluminium alloys in view of the increase in overall efficiency and the requirement in strength [15, 17]. Aluminium alloys that respond to strengthening by heat treatment are covered by three series, namely, 2xxx (Al–Cu, Al–Cu–Mg), 6xxx (Al–Mg–Si) and 7xxx (Al–Zn–Mg, Al–Cu–Mg–Cu). These alloys offer excellent strength-to-weight ratio and reasonably good corrosion resistance [20]. Out of these aluminium alloys, 2xxx and 7xxx series alloys are generally classified as non-weldable because of poor solidification microstructure and porosity in the fusion zone. In addition, the loss in mechanical properties and distortion of the components during fusion welding techniques as compared to the base material is very significant [16]. Fusion welding of age hardenable aluminium alloys is inappropriate for many structural components due to low joint strength. Riveting has been the most popular method for joining such alloys. The use of huge quantity of rivets in airplane structures and panels leads to increase in weight. It also adds the complexity of stress concentration in riveted area along with possibility of corrosion, which may lead to failure in function [1].

A solid-state welding method developed in the early 1990s generated significant interest because of a whole range of advantages over conventional techniques [5]. The problems associated with joining of high-strength aluminium alloys were solved with the application of solid-state joining method. This new technique, Friction Stir Welding (FSW), can weld two sheets in the solid phase without melting the parent metal. Thomas et al. [28] invented this process at The Welding Institute of United Kingdom in 1991. FSW is an energy-efficient, environment-friendly, green solid-state welding process. There have been widespread benefits in joining high-strength aluminium alloys for aerospace, marine and automotive industries by using FSW. This solid-state joining technique is very popular as 'Green Technology' and 'Environment-friendly process' [2, 14, 16].

During FSW, tight clamping arrangements of the plates are essential. A rotating tool is plunged in joint line at one end until it reaches to predetermined depth. Tool rotates at a constant speed for small duration of time known as dwell period. The continuous rotation of tool leads to softening of material around the pin and below the shoulder surface. As the tool moves along joint line, softened metal is forced to flow from leading edge to the trailing edge of the pin; it then cools to form the solid-state joint. Because of the extreme level of plastic deformation, a very fine, dynamically recrystallized microstructure is obtained in the weld zones [16, 29]. The joint is stronger and comparatively defect free than fusion welding. The heat generated during FSW in comparison with fusion welding methods causes minimum residual stresses and associated defects.

Generally, during FSW, the peak temperature reaches the solutionizing temperature of age hardenable aluminium alloys. Its range depends on material composition, type, initial temper and process parameters [7]. The FSW exhibits a temperature distribution from weld centre towards base metal (BM). Due to the distribution of temperature and its gradient, a variety of precipitate evolution is expected. This leads to a variation in the microstructure across various regions. The microstructural characteristics such as grain size, precipitate type, size, number density and distribution and dislocation density control the mechanical properties and corrosion behaviour [10, 13, 16]. It is possible to detect the variations in microstructure by measuring the electrical conductivity (EC) as non-destructive testing (NDT) technique [24, 25, 30].

FSW achieves solid-state joining by local frictional heat and plastic flow, which results in a change in the local microstructure of age hardenable aluminium alloys [31]. Normally, the FSW joint section is divided into four zones: nugget zone (NZ) or dynamically recrystallised zone (DXZ), thermo-mechanically affected zone (TMAZ), heat-affected zone (HAZ) and unaffected base metal (BM) [23, 27]. The weld sections of 2xxx, 6xxx and 7xxx series of aluminium alloys are susceptible to microstructural changes during FSW due to their ageing characteristics. A more pronounced effect is seen eventually in 7xxx series aluminium alloys through the improvement in mechanical properties and reduction in corrosion resistance with respect to time. The most feasible solution to stabilize the response is to apply post-weld heat treatment (PWHT) to the weld to restore the corrosion resistance with minimum loss in strength. The modification of size, shape and distribution of strengthening particles plays a vital role in restoring the mechanical properties as well as corrosion resistance. It is quite possible by adapting PWHT to FSW joint [3, 8, 12, 32] .

The proper implementation of PWHT to FSW joints of age hardenable aluminium alloys is important for achieving favourable microstructure, because it influences the mechanical properties as well as corrosion behaviour [11]. Kumar et al. [12] studied the effect of PHWTs, viz. peak ageing (T6) and retrogression and re-ageing (RRA) on the microstructure, mechanical properties and pitting corrosion. The observations revealed through the investigation done by Kumar et al. [12] indicate an increase in hardness and tensile strength of joint in T6 condition with a reduction in corrosion resistance. The pitting corrosion resistance noted an improvement in RRA condition with minimum loss of weld strength. Kumar et al. [12] noted that microstructure in T6 condition has relatively coarse and closely spaced precipitates along the grain boundaries and the fine precipitates within the grains resulting in high hardness and strength. On the other hand, in RRA condition, the grain boundary precipitates are discontinuous and coarser, which promote the equilibrium phase in the grains and subgrain boundaries. Due to discontinuous grain boundary precipitates with large spacing, no continuous chain is formed for corrosion to take place. Cerri [3] studied the effect of PWHT at 200, 300, 400 and 450 °C for a time duration of 0.5–6 h to investigate microstructure and mechanical properties evolution during FSW of AA7075. The microstructural analysis revealed a progressive change in grain size and observed abnormal grain growth in the stir zone. Sivaraj et al. [26] studied the effect of post-weld heat treatment on tensile properties and microstructural characteristics of FSWed armour grade AA7075. The welded joints were subjected to two different heat treatment cycles, namely, solution treatment followed by artificial ageing (STA) and artificial ageing (AA).

The AA treatment performed on FSW joint minimized yield strength and tensile strength of as-welded (AW) joint. The STA treatment increased yield strength and tensile strength, resulting in increasing joint efficiency in comparison to AW joint.

Previous research shows that FSW reduces exfoliation corrosion resistance of AA 7075 when the joint left in as-welded condition or provided PWHT like artificial ageing. Petter et al. [19] patented heat treatment cycle for FSWed 7X50 aluminium alloys, and represented that this methodology improved corrosion resistance. The heat treatment method described in five stages includes solution heat treating, quenching, stabilizing in ambient air at room temperature, first ageing and second ageing. The post-weld solution heat treatment reduces material degradation as it homogenizes the material affected by FSW. It also significantly limits large grain growth at the face and root side of the weldment. Reduction in residual stress induced during FSW is another benefit that occurred through PWHT. This was the key motivation to consider high-temperature solution treatment, quenching, stabilization and double ageing as PWHT for FSW joints of AA 7075, which is named herein as stabilization and double ageing SDA.

The literature review clearly shows that PWHT can efficiently modify the microstructure, which leads to improvement in the mechanical properties and resistance to corrosion of FSW joints. There are many methods such as T6, T7, RRA, AA and STA proposed for improving the mechanical properties and corrosion resistance of AA7075 FSW joints. The corrosion resistance of FSW joints, which are heat treated to RRA condition, is better than other treatment conditions without considerable loss in strength. Hence, the PWHTs, RRA and SDA were selected to study the microstructure, mechanical properties, electrical conductivity and corrosion resistance of FSW joints. In the present work, the PWHT, namely, RRA and SDA are carried out to investigate the effect on microstructure, mechanical properties, electrical conductivity and corrosion behaviour of AA7075-T6 FSW joints. The main contributions of this chapter are as follows. First, two different PWHTs such as RRA and SDA are adopted to assess its effect on corrosion resistance and mechanical properties of FSW joints. Second, a SDA PWHT compared with RRA method based on the measurements of tensile strength, hardness, electrical conductivity and exfoliation corrosion rate. The outcome clearly shows that there is a response of this alloy joints to PWHT with respect to various properties under certain treatments. The rest of this chapter is organized as follows. Section 4.2 describes the experimental work and methodology, Sect. 4.3 deals with results comprising microstructural characterization, effect on hardness, electrical conductivity, and the results are discussed in Sect. 4.4. Section 4.5 concludes the chapter.

4.2   Experimental Work and Methodology

Rolled aluminium alloy plates of AA 7075-T6 are considered as a base metal (BM) for this investigation. The plates were sectioned with size 300 mm × 60 mm × 6 mm and rigidly clamped for processing FSW joints. The chemical

Table 4.1 Chemical composition (wt%) of base metal

Material	Al	Si	Cu	Mg	Cr	Mn	Ti	Fe	Zr	Zn
AA7075	Bal.	0.21	1.32	2.30	0.18	0.10	0.12	0.37	0.01	5.21

Table 4.2 Mechanical properties of the base metal

Material	UTS (MPa)	YS (MPa)	% Elongation	Hardness (HV)
AA7075	582	469	16	196

composition in weight percentage and mechanical properties of a base metal are described in Tables 4.1 and 4.2, respectively.

Butt joints of AA 7075-T6 plates produced along the longitudinal direction using a computer numerically controlled (CNC) milling machine. The FSW tool of H13 material has a shoulder with 15 mm diameter and a 5.4 mm height, threaded conical probe with 6 mm at top and 5 mm at tip. The process parameters were optimized using a series of trial runs to obtain macro level defect-free joints. The tool rotation speed of 500 RPM and welding speed of 60 mm/min were used to fabricate these joints.

In order to study the effect of PWHT on the microstructure and mechanical properties, RRA and SDA treatments were applied to FSW joint. The RRA treatment was carried out in four steps, solution treatment at 480 °C for 30 min, water quenching, retrogression at 180 °C for a soaking period of 30 min and the last stage as re-ageing at 120 °C for 24 h. The schematic representation of RRA treatment described in Fig. 4.1 shows all the four stages. The SDA treatment was applied in five steps, solution heat treatment at 480 °C for 30 min, quenching, stabilization at room temperature for 96 h, first-stage ageing at 120 °C for 5 h and second-stage ageing at 160 °C for a long duration soaking period of 27 h and then air cooled. The schematic representation of SDA treatment described in Fig. 4.2 shows all the above five stages.

Metallographic specimens were prepared by standard metallographic technique. Keller's reagent (150 mL H_2O, 3 mL HNO_3 and 6 mL HF) was used for etching to reveal the grain structure of different weld zones. Optical microscope (Make: ZEISS) and SEM (JEOL 6380A, Japan) were used to capture the image of the specimen. The specimens were sectioned from NZ, TMAZ, HAZ and BM regions of weldment for metallurgical characterization.

Hardness testing was carried out using Vickers hardness tester (Make: Omni Tech) with a load of 500 g. f. and dwell time of 10 s. At regular intervals of 5 mm from the centerline of the weld, on both sides of the weld, the hardness was tested. Tensile test specimens are prepared using ASTM-E8 standard and testing was performed on a computer controlled Universal Testing Machine at a crosshead speed of 3 mm/min. The ultimate tensile strength (UTS), yield strength (YS) and percentage of elongation (% E) were evaluated. The measurement of electrical conductivity at different zones of weldment was carried out within 1% accuracy by

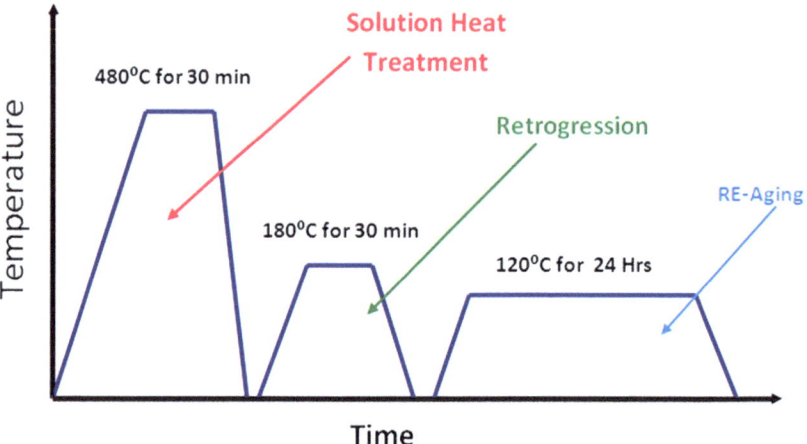

Fig. 4.1 Schematic representation of RRA as PWHT for AA 7075-T6 FSW joints

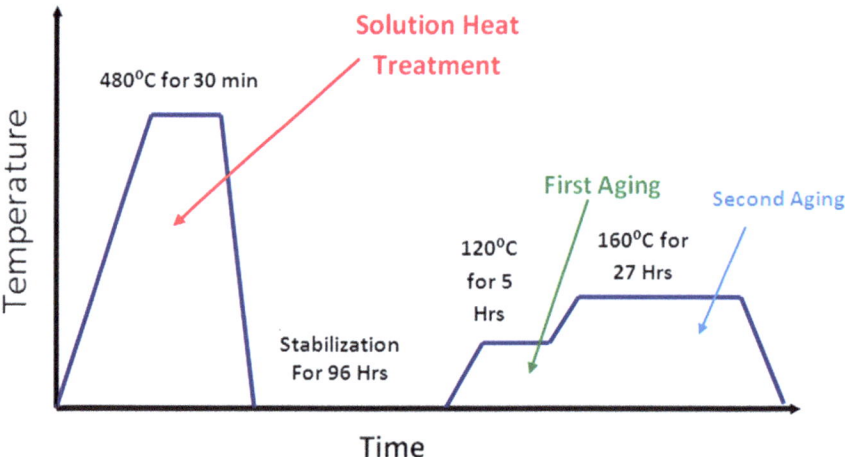

Fig. 4.2 Schematic representation of SDA as PWHT for AA 7075-T6 FSW joints

using electrical conductivity meter in % IACS (International Annealed Cooper Standard). It is a standard practice to express the conductivity of age hardenable aluminium alloys as % IACS as per ASTM E1004-09.

Exfoliation test was performed as per ASTM G34-01 (Exfoliation Corrosion Susceptibly in 7XXX Series aluminium alloys). The EXCO test solution was prepared using distilled water with reagent-grade sodium chloride (NaCl), potassium nitrate (KNO_3) and nitric acid (HNO_3). The mixture of NaCl (234 g) and KNO_3 (50 g) was dissolved in distilled water. To this solution, 6.3 mL of concentrated HNO_3 was added. Later, the solution was diluted to 1 L and the pH of 4.0

(a)   (b)

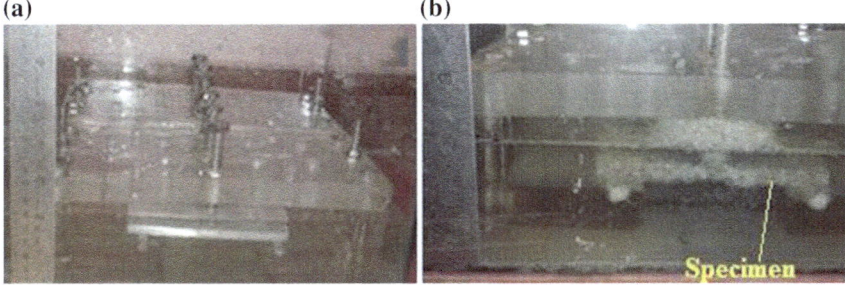

Fig. 4.3 Exfoliation corrosion test setup, **a** specimen clamped and supported, **b** specimens in immersed condition after 24 h showing corrosion-prone area

was maintained. The EXCO solution composition was represented as NaCl (4.0 M), KNO_3 (0.5 M) and HNO_3 (0.1 M). In the composition, 'M' represents molarity corrosion rate calculated based on mass loss method. All the samples (including as-welded, RRA treated and SDA treated) were exposed to an EXCO solution for 48 h. After proper cleaning, all the test coupons were weighted. The weight loss was determined as the difference between the initial weight of the specimen prior to immersion and its weight after removal of corrosion product. The corrosion rate for each coupon was calculated based on mass loss method according to ASTM-G1 standard using Eq. (4.1), which is described as follows. Figure 4.3 represents the experimental setup for immersion exfoliation corrosion.

The unit of corrosion rate (CR) is described in mm/year.

$$\mathrm{CR} = (K \times W)/(A \times t \times \delta), \tag{4.1}$$

where

K constant (8.76×10^4 for CR in mm/year);
W weight loss in gm;
δ density of aluminium alloy in gm/cm^3;
t exposure time in hours;
A total surface area of the sample in cm^2.

4.3 Results

4.3.1 Tensile Properties

The tensile test was carried out for the FSW joints in AW, SDA and RRA conditions along with BM. In each condition, three samples of each condition were tested and the average values are presented in Table 4.3. The ultimate tensile strength and yield strength of the un-welded parent metal are 582 and 502 MPa,

Table 4.3 Summary of the transverse weld tensile properties of AA 7075-T6 joints

Condition	Tensile strength (MPa)	Yield strength (MPa)	Elongation (%)	Joint efficiency (%)
BM	582	469	16	–
AW	448	390	7	77
RRA	581	563	11	99
SDA	541	491	14	93

respectively, with an elongation of 16%. However, the FSW joint exhibits lower ultimate and yield strength of 449 and 390 MPa, respectively, with an elongation of 7%. This suggests that FSW causes a huge reduction in tensile strength (23%) of AA7075-T6 alloys, similar kind of results were reported by other researchers [21, 22]. The SDA treatment performed on the FSW joint increases the tensile strength and yield strength to 541 and 491 MPa, respectively, resulting in an increase in the joint efficiency by 16% in comparison to AW joint. The SDA treatment also causes an increase in elongation by 7% in comparison with the AW joint. The RRA treatment increases the tensile strength and yield strength to 582 and 563 MPa, respectively, resulting in an increase in the joint efficiency by 23% in comparison to AW joint. It caused an increase in elongation by 4% for the AW joint.

4.3.2 Hardness Test

The hardness at various locations from the weld centre was measured using a Vickers microhardness tester. The hardness was measured on the transverse cross section of the joints. The hardness profiles of the cross section of the welded joints with RRA treatment, SDA treatment; AW condition and BM are presented in Fig 4.4. The hardness measurements were taken at the interval of 5 mm from the centre of weld zone. Figure 4.4 also represents the weld section (30 mm) indicating different zones as NZ, TMAZ and HAZ. The influence of PWHT clearly shows a variation of hardness values. In AW condition, the minimum hardness found near the boundary of TMAZ and HAZ on both sides—advancing side (AS) as well as retreating side (RS)—are 127 HV and 143 HV, respectively. The nature of profile of hardness of AW sample shows typical nearly 'W' form. The average hardness value at NZ is 161 HV in AW condition that is less than the hardness of BM. The profile clearly represents that hardness is increased in almost all zones of joint after RRA treatment and it is higher as compared to SDA treatment and AW samples. After PWHT-RRA, the hardness increased more than BM sample in NZ of joint.

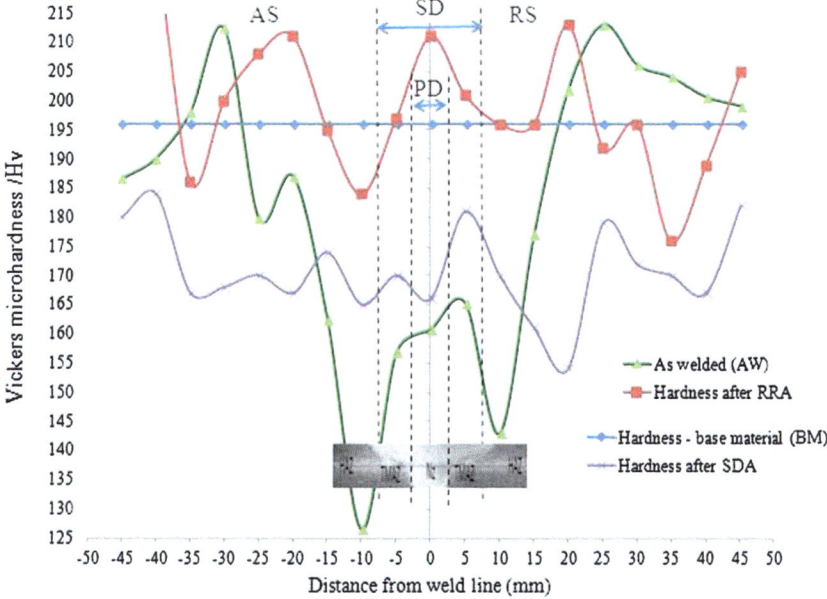

(AS-advancing side, RS-retreating side, SDshoulder diameter and PD--pin diameter)

Fig. 4.4 Hardness profiles across the FSW joint of AA 7075-T6

4.3.3 *Electrical Conductivity Measurements*

Electrical conductivity (% IACS) was measured after each heat treatment and compared to BM and AW condition. Figure 4.5 represents that the electrical conductivity of RRA- and SDA-treated FSW joints have a uniform response as compared to AW condition. The weld section (30 mm) indicating different zones as NZ, TMAZ and HAZ correlated with variations in electrical conductivity.

As shown in Fig. 4.5, the electrical conductivity rose to 37% IACS and remains constant throughout the section of joint after PWHTs (RRA and SDA). In AW condition, the electrical conductivity profile shows 'M' pattern, which is opposite to hardness profile of AW condition shown in Fig. 4.4.

4.3.4 *Microstructure*

Figure 4.5 represents the optical micrographs of BM NZ and TMAZ of AW, RRA, SDA conditions. In the base metal, equiaxed grains are oriented along the rolling direction, as shown in Fig. 4.5a. The NZ region of AW joint reveals a fine-grain structure due to the dynamic recrystallization during FSW represented in Fig. 4.6b.

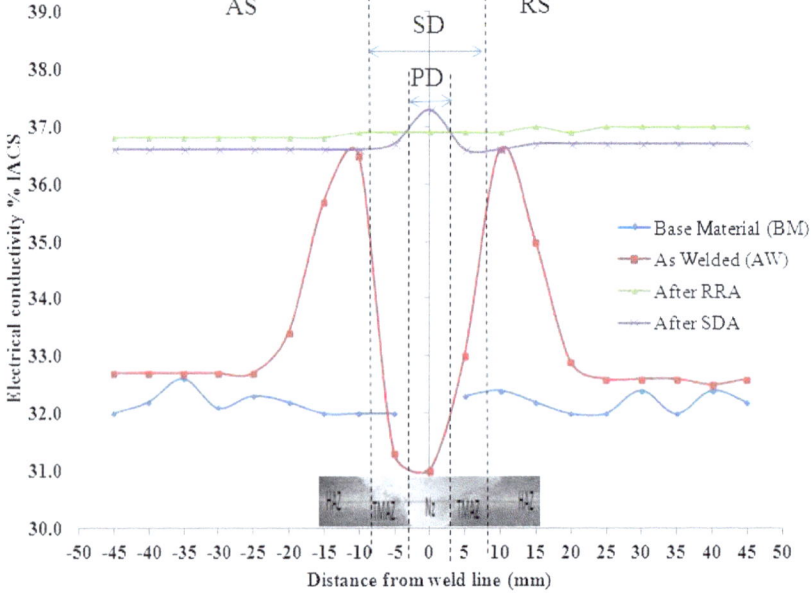

(AS-advancing side, RS-retreatingside, SD-shoulder diameter and PD-pin diameter)

Fig. 4.5 Electrical conductivity profiles across the FSW joint of AA 7075-T6

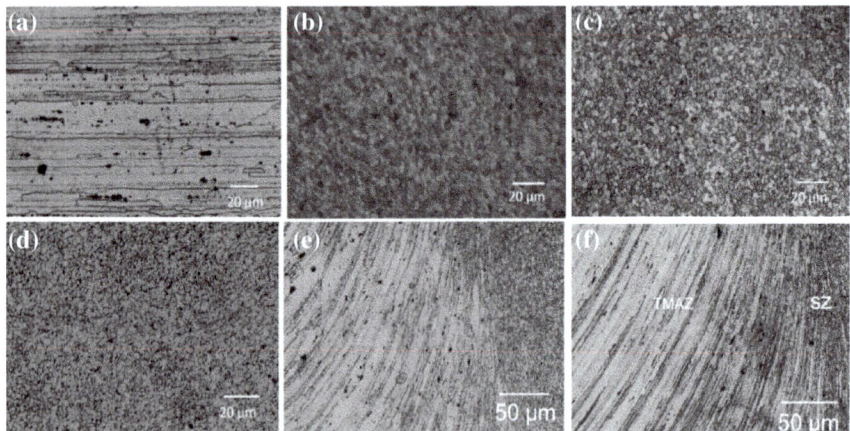

Fig. 4.6 Microstructure of FSW: **a** base metal, **b** as-welded SZ, **c** PWHT-RRA SZ, **d** PWHT-SDA SZ, **e** and **f** transition zone showing SZ and TMAZ at RRA and SDA conditions, respectively

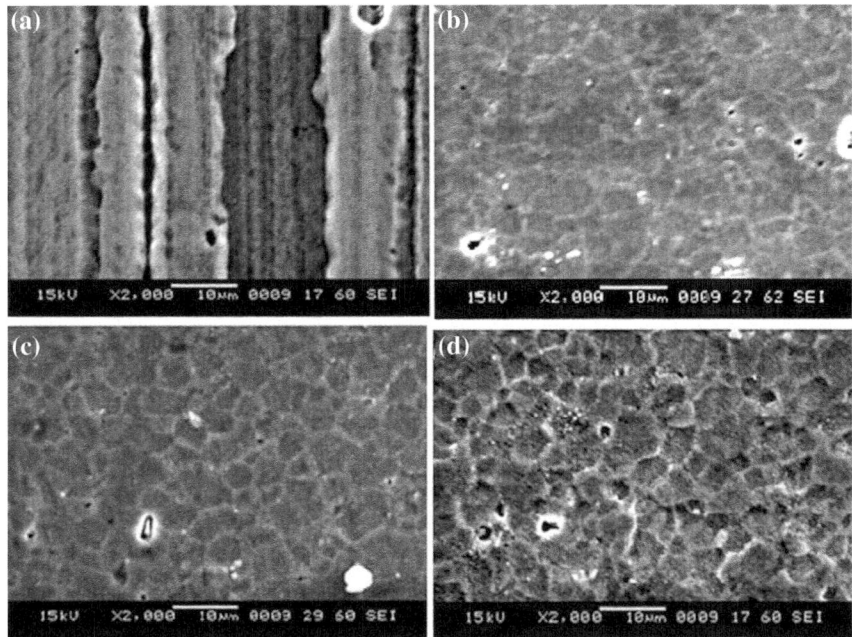

Fig. 4.7 SEM images of the surface after corrosion testing, **a** base metal, **b** SDA-treated nugget zone, **c** RRA-treated nugget zone, **d** AW nugget zone

Figure 4.6c, d represents the NZ of RRA- and SDA-treated FSW joints, respectively, indicating grain structure different from AW joint section. The region adjacent to the NZ, i.e. TMAZ is characterized by a highly deformed structure as shown in Fig. 4.6e, f. The SEM images of the surface of FSW joints at different conditions such as AW, RRA treated and SDA treated are represented in Fig. 4.7.

Figure 4.8 represents the EDX image of nugget zone of FSW joint with SDA as PWHT. It clearly shows the presence of intermetallic constituents of Mg, Cu, Fe and Zn. Further, the specific intermetallic or precipitates can be identified using TEM analysis. TEM analysis is out of the scope of this present work. The characteristic microstructural constituents formed following retrogression and re-ageing condition consist of grain boundary precipitate (GP) zones and $MgZn_2$ precipitates.

4.3.5 Exfoliation Corrosion

Exfoliation corrosion test was performed according to ASTM G34 exfoliation corrosion susceptibly in 7XXX Series aluminium alloys (EXCO Test) for BM, AW, RRA and SDA conditions by exposing specimens to an acidic solution for 48 h.

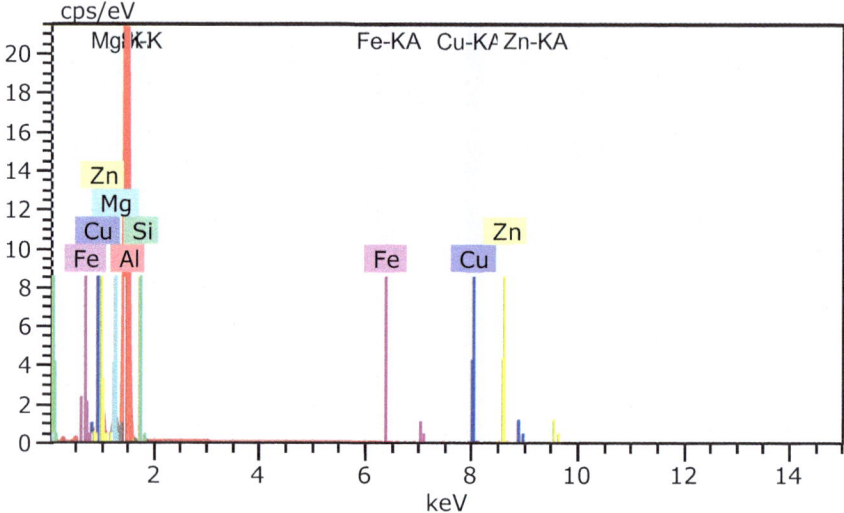

Fig. 4.8 EDX image of nugget zone of FSW joint post-weld heat treated as SDA

	S. No.	Material condition	Corrosion rate (mm/year)
Table 4.4 Corrosion rate of FSW joint in AW, RRA and SDA conditions	1	BM	14.62
	2	AW	8.69
	3	RRA	3.48
	4	SDA	2.32

At least two samples were prepared for each condition tested. Table 4.4 represents corrosion rate results. The corrosion rate of SDA-treated samples was found less than 2.32 mm/year as compared to RRA-treated and AW condition samples.

Figure 4.9 describes the severity of corrosion attack at different zones of FSW joint. The samples were kept in corrosive media for 48 h. After cleaning the samples, the macrostructure was observed. It shows different corrosion behaviours at different zones of FSW joints. As shown in Fig. 4.9a, maximum corrosion pits are observed at the top surface in HAZ zone. Whereas at the bottom side, the corrosion pits are observed at root side of joints. The boundary of HAZ and TMAZ is more prone to corrosion attack.

Fig. 4.9 FSW joint samples after exfoliation test NaCl (4.0 M), KN0$_3$ (0.5 M) and HN0$_3$ (0.1 M), for 48 h, **a** representation of top side of FSW joint, **b** representation of root side of FSW joint and **c** representation of corrosion attack at HAZ and TMAZ cross-sectional view

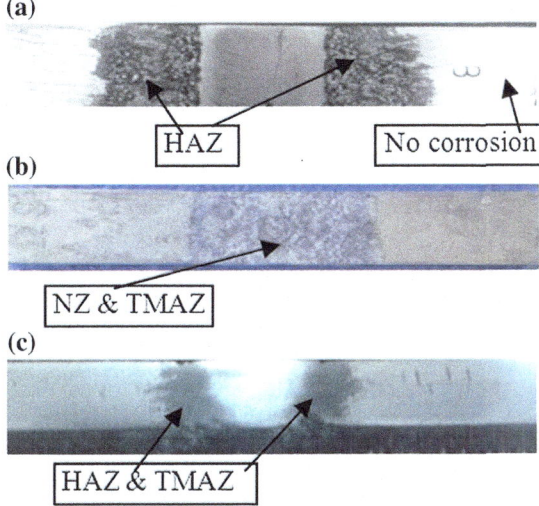

4.4   Discussion

Although FSW is generally considered as a simple process, the joint formation during FSW is complex. During this process, thermal and mechanical deformation conditions vary across the weld region, resulting in varied and different microstructures (Fig. 4.6). Since the joining of plates is at solid state, the stirring action of tool leads to fine grains in the weld centre as shown in Fig. 4.6b–d. The TMAZ adjacent to SZ typically has a lower level of deformation and temperature in comparison to the SZ. The microstructure pattern at the transition section of SZ and TMAZ (Fig. 4.6e, f) is different and leads to variation in hardness and electrical conductivity as shown in Figs. 4.4 and 4.5, respectively. The modification in microstructure induced by FSW is reflected in tensile strength, ductility, hardness and electrical conductivity. Subsequently, it is also reflected in corrosion behaviour of as-welded and post-weld heat-treated samples. The tensile strength and ductility of welded joints in different PWHT conditions and AW condition are described in Table 4.3. It indicates that AW joint has lower tensile strength than BM. The failure of AW joint occurs at TMAZ region, which has lowest hardness value (127 HV). The boundary between TMAZ and HAZ on advancing side is the weakest zone. The RRA treatment to FSW joint increases the tensile strength of the joint along with an increase in hardness, electrical conductivity and corrosion resistance. The high strength of RRA microstructure is mainly due to GP zones, but the contribution of strengthening precipitates to the strength was reported to be higher [9].

The minimum hardness (127 HV) on advancing side in HAZ region of AW sample is attributed to the growth and over-ageing of precipitates. This effect is because the HAZ region is affected only by the temperature gradient. In this region, there is no thermo-mechanical work. SZ has higher hardness (Fig. 4.3) values than TMAZ region and HAZ region but less as compared to BM hardness and this is due to dynamic recrystallization. The phenomenon of dynamic recrystallization and its effect on hardness after FSW is reported in earlier research [4]. The hardness in SZ and TMAZ (184 HV) increased during post-weld RRA treatment. The post-weld SDA joint represents high tensile strength than AW joints, but lower than the post-weld RRA-treated samples. During FSW, the stir zone, which has undergone dissolution of fine precipitates gets re-precipitated during RRA and SDA treatments; thereby, the hardness is increased in this region [33]. The tensile test results show that the RRA and SDA treatments have improved the tensile strength of the FSW joints with a drastic improvement in hardness values across the FSW joint (Fig. 4.4), and similar results were observed by other researchers [6].

The electrical conductivity of nonferrous materials depends on the electron mobility, the crystalline structure of existing phases as well as crystal defect content. It is a quick method to evaluate the joint integrity and temper condition of material. The measurement of electronic mobility and resistance to flow of electrons describes the electrical conductivity. The electrical conductivity that increased after post-weld RRA and SDA treatments shows uniformity along the joint (Fig. 4.5). The electrical conductivity of RRA joints is slightly higher than SDA. The uniformity in electrical conductivity is attributed to uniform distribution of precipitates caused due to post-weld heat treatment. This uniform distribution of precipitates is the cause of improvement in exfoliation corrosion resistance by post-weld heat treatments of RRA and SDA. During FSW, heat is generated due to friction and because of heat transfer and distribution, the material experiences different temperatures at different weld zones resulting in changes in microstructure. The distribution of precipitates affects the hardness, corrosion behaviour and electrical conductivity [18]. This investigation has revealed an additional benefit of SDA PWHT in terms of improvement in the corrosion resistance along with tensile properties.

4.5   Conclusions

The PWHT, namely, RRA and SDA were carried out to investigate the effect on microstructure, hardness, tensile strength, electrical conductivity and corrosion behaviour of AA7075-T6 FSW joints. The following conclusions were drawn from the results obtained in the present work.

(1) After FSW of AA 7075-T6, the tensile strength was reduced by 23% in comparison to tensile strength of base metal before welding. The tensile strength has recovered after PWHTs i. e., 22% by RRA and 16% SDA.
(2) Fine-grained microstructure with a uniform distribution of fine precipitates in weld nugget was found to be an important factor responsible for improved tensile strength due to RRA and SDA treatments.

(3) The SDA treatment applied in this investigation was found beneficial to exfoliation corrosion resistance of FSW joint, which is superior to the RRA treatment.

(4) SDA treatment can be considered as PWHT for FSW joint of AA 7075-T6 alternative for RRA treatment.

Acknowledgments The authors are grateful to Dr. Dilip Peshwe, Professor, Department of Metallurgy and Materials Engineering, VNIT, Nagpur and Center of Excellence of Metallurgy and Materials Engineering, MIT Aurangabad for providing support to carry out the characterization.

References

1. Bahemmat, P., Haghpanahi, M., Besharati, M.K., Ashanizadeh, S., Rezaei, H.: Study on mechanical, micro and microstructural characteristics of dissimilar friction stir welding of AA6061-T6 and AA 7075-T6. Proc. Inst. Mech. Eng. Part B: J. Eng. Manuf. **224**, 1854–1865 (2010)
2. Cam, G., Mistikoglu, S.: Recent developments in friction sir welding of Al alloys. J. Mater. Eng. Perform. **23**(6), 1936–1953 (2014)
3. Cerri, E.: Effect of post-welding heat treatments on mechanical properties of double lap FSW joints in high strength aluminium alloys. Metall. Sci. Technol. **29–1**, 32–39 (2011)
4. Elangovan, K., Balasubramanian, V.: Influences of post-weld heat treatment on tensile properties of friction stir-welded AA6061 aluminium alloy joints. Mater. Charact. **59**, 1168–1177 (2008)
5. Enomoto, M.: Friction stir welding: research and industrial applications. Weld. Int. **17**(5), 341–345 (2003)
6. Guo-sheng, P., Kang-hua, C., Song-yi, C., Hua-chan, F.: Influence of dual retrogression and re-aging temper on microstructure, strength and exfoliation corrosion behavior of Al-Zn-Mg-Cu alloy. Trans. Nonferrous Met. Soc. China **22**, 803–809 (2012)
7. Hassan, K.A.A., Prangnell, P.B., Norman, A.F., Price, D.A., Williams, S.W.: Effect of welding parameters on nugget zone microstructure and properties in high strength aluminium alloy friction stir welds. Sci. Technol. Weld. Joining **8**(4), 257–268 (2003)
8. Ipekoglu, G., Erim, S., Cam, G.: Investigation of the effect of temper condition and post weld heat treatment on the microstructure and mechanical properties of friction stir butt welded AA 7075 Al alloy plates. Int. J. Adv. Manuf. Technol. **70**, 201–213 (2015)
9. Isadare, A.D., Aremo, B., Adeoye, M.O., Olawale, O.J., Shittu, M.D.: Effect of heat treatment on some mechanical properties of 7075 aluminium alloy. Mater. Res. **16**(1), 190–194 (2013)
10. Jasthi, B.K., Klinckman, E., Curtis, T., Widener, C., West, M., Ruokolainen, R.B., Dasgupta, A.: Effect of post-weld aging on the corrosion resistance and mechanical properties of friction stir welded aluminium alloys 7475-T73. In: Friction Stir Welding and Processing VIITMS, The Minerals, Metals & Materials Society, pp. 225–234 (2013)
11. Kumar, P.V., Reddy, M.G., Rao, S.K.: Effect of post weld heat treatments on mechanical and stress corrosion cracking behaviour of AA 7075 friction stir welds. Int. J. u- e-Serv. Sci. Technol. **7**(4), 251–262 (2014)
12. Kumar, P.V., Reddy, M.G., Rao, S.K.: Microstructure, mechanical and corrosion behaviour of high strength AA7075 aluminium alloy friction stir welds—effect of post weld heat treatment. Def. Technol. **11**, 362–369 (2015)
13. Lumsden, J.B., Mahoney, M.W., Pollock, G., Rhodes, C.G.: Intergranular corrosion following friction stir welding of aluminum alloy 7075-T651. Corrosion **55**(12), 1127–1135 (1999)

14. Mahoney, M.W., Rohdes, C.G., Flintoff, J.G., Bingel, W.H., Spurling, R.A.: properties of friction-stir welded 7075 T651 aluminum. Metall. Mater. Trans. A **29**(7), 1955–1964 (1998)

15. Miller, W.S., Zhuang, L., Bottema, J., Wittebrood, A.J., De Smet, P., Haszler, A., Vieregge, A.: Recent development in aluminum alloys for the automotive industry. Mater. Sci. Eng. A **280**, 37–49 (2000)

16. Mishra, R.S., Ma, Z.Y.: Friction stir welding and processing. Mater. Sci. Eng. R **50**, 1–78 (2005)

17. Mohammadi-pour, M., Khodabandeh, A., Mohammadi-pour, S., Paidar, M.: Microstructure and mechanical properties of joints welded by friction-stir welding in aluminum alloy 7075-T6 plates for aerospace application. Rare Met. 1–9 (2016). https://doi.org/10.1007/s12598-016-0692-9

18. Patil, S., Gogte, C.L.: On thermal effects in the weld zone of friction stir welded joint of age hardenable AA 7075 alloy. In: 7th Asia Pacific IIW International Congress, Singapore, ICRA-2013-SING.235, IIW (2013)

19. Petter, G.E., Figert, J.D., Rybicki, D.J., Burns, T.H.: Heat treatment of friction stir welded 7X50 aluminium. U.S. Patent 6 802 441 B1, 12 Oct 2004 (2004)

20. Polmear, I.: Light Alloys: From Traditional Alloys to Nanocrystals, 4th edn. Elsevier, Amsterdam (2006)

21. Rafi, K.H., Janaki Ram, G.D., Phanikumar, G., Rao, P.K.: Microstructure and tensile properties of friction stir welded aluminium alloy AA7075-T6. Mater. Design **31**, 2375–2380 (2010)

22. Rajakumar, S., Muralidharan, C., Balasubramanian, V.: Influence of friction stir welding process and tool parameters on strength properties of AA7075-T6 aluminium alloy joints. Mater. Design 1–15 (2010)

23. Rhodes, C.G., Mahoney, M.W., Bingel, W.H., Spurling, R.A., Bampton, W.H.: Effects of friction stir welding on microstructure of 7075 aluminium. Scripta Mater. **36**(1), 69–75 (1997)

24. Santos, T.G., Miranda, R.M., Vilaca, P., Teixeira, J.P.: Modification of electrical conductivity by friction stir processing of aluminum alloys. Int. J. Adv. Manuf. Technol. **57**, 511–519 (2011)

25. Santos, T.G., Miranda, R.M., Vilaca, P., Teixeira, J.P., dos Santos, J.: Microstructural mapping of friction stir welding AA 7075-T6 and AlMgSc alloys using electrical conductivity. Sci. Technol. Weld. Joining **16**(7), 630–635 (2011)

26. Sivaraj, P., Kanagarajan, D., Balasubramanian, V.: Effect of post weld heat treatments on tensile properties and microstructure characteristics of friction stir welded armour grade AA 7075-T651 aluminium alloy. Def. Technol. **10**, 1–8 (2014)

27. Su, J.Q., Nelson, T.W., Mishra, R.S., Mahoney, M.W.: Microstructural investigation of friction stir welded 7050-T651 aluminum. Acta Mater. **51**, 713–719 (2005)

28. Thomas, W.M., Nicholas, E.D., Needham, J.C., Smith, P.J., Kallee, S.W., Dewas, C.: Friction stir welding. UK Patent Publication, GB 2.306 266 (1995)

29. Thomas, W.M., Staines, D.G., Norris, I.M., de Frias, R.: Friction stir welding tools and developments. Weld. World **47**(11–12), 10–17 (2003)

30. Tsai, T.C., Chuang, T.H.: Relationship between electrical conductivity and stress corrosion cracking susceptibility of Al 7075 and Al 7475 alloys. Corrosion **52**(6), 414–416 (1996)

31. Venugopal, T., Rao, S.K., Rao, P.K.: Studies on friction stir welded AA 7075 aluminium alloys. Trans. Indian Inst. Met. **57**(6), 659–663 (2004)

32. Yani, C., Sayer, S., Ertugrul, O., Pakdil, M.: Effect of post-weld aging on the mechanical and microstructural properties of friction stir welded aluminium alloy 7075. Arch. Mater. Sci. Eng. **34**(2), 105–109 (2008)

33. Zaid, H.R., Hatab, A.M., Ibrahim, A.M.A.: Properties enhancement of Al-Zn-Mg alloy by retrogression and re-aging heat treatment. J. Min. Metall. B: Metall. **47**(1), 31–35 (2011)

Chapter 5
Influence of Tool Plunge Depth During Friction Stir Spot Welding of AA5052-H32/HDPE/AA5052-H32 Sandwich Sheets

Pritam Kumar Rana, R. Ganesh Narayanan and Satish V. Kailas

Abstract Metal/polymer/metal multi-layered materials have shown promising properties because of lightweight characteristics in automotive industries. Joining of these materials is difficult by conventional methods due to large difference in their physical and chemical properties. In the present work Friction Stir Spot Welding (FSSW) of AA5052-H32/HDPE/AA5052-H32 sandwich sheet is done. The objective is to analyse the influence of tool plunge depth on the joint behaviour. This is accomplished through joint characterization by evaluating mechanical performance, hook and flash formation, grain size, temperature measurement, and hardness distribution. Lap shear test, cross-tension test, peel test, and uni-axial tensile tests are conducted. A comparison between bimetallic and sandwich sheet has also been done. First, for joining sandwich sheets, the optimum plunge depth is 3.6 mm and greater. Adequate joint strength and extension at failure are obtained in this range. The joint strength does not depend on hook geometry, rather it depends on bond width and joint hardness. Second, though the joint strength of sandwich sheets is reduced as compared to bimetallic, the flash formed is minimised in sandwich sheets. The deformed material gets accommodated in the core layer region to reduce the flash formation. Finer grains are seen in sandwich sheet due to lesser peak temperature. Nugget pull out failure is commonly seen after testing and is independent of test method and the plunge depth.

Keywords Sandwich sheets · Spot welding · Joint strength · Hook
Temperature · Mechanical tests

P. K. Rana · R. Ganesh Narayanan (✉)
Department of Mechanical Engineering, IIT Guwahati, Guwahati 781039, India
e-mail: ganu@iitg.ernet.in

P. K. Rana
e-mail: r.pritam@iitg.ernet.in

S. V. Kailas
Department of Mechanical Engineering, IISc Bangalore, Bengaluru 560012, India
e-mail: satvk@iisc.ac.in

© Springer Nature Singapore Pte Ltd. 2019
U. S. Dixit and R. G. Narayanan (eds.), *Strengthening and Joining by Plastic Deformation*, Lecture Notes on Multidisciplinary Industrial Engineering,
https://doi.org/10.1007/978-981-13-0378-4_5

5.2 Experimental Procedure

In this section, the FSSW trials, sample preparation, mechanical testing, joint characterization, and temperature measurement procedures are described.

5.2.1 FSSW and Sample Preparation

The present work is executed on 2 mm thick AA5052-H32 sheets and 1 mm thick (HDPE) sheet. Table 5.1 shows the mechanical properties of AA5052-H32 and HDPE. ASTM B557 M-15 and ASTM D 638-14 standards are used to evaluate tensile properties of AA5052-H32 and HDPE respectively.

All the tensile tests are carried out on a 200 kN universal testing machine operated by hydraulic loading (Model: UTE 20, Make: FIE) at a strain rate of 0.02/ min. and at room temperature. Comparative stress–strain behaviour of AA5052-H32 and HDPE sheet is shown in Fig. 5.2. The tensile testing of HDPE sheet is stopped manually because it has not failed after a considerable amount of deformation.

The chemical composition of AA5052-H32 alloy obtained from energy dispersive X-ray analysis is given in Table 5.2.

AA5052-H32 sheets are placed as the outer layers and the HDPE sheet is placed in between to prepare the sandwich sheets. The tool used for making FSSW joint is made up of H13 tool steel. The tool has simple profile with flat shoulder and straight cylindrical pin. The dimensions of different tool geometries are shown in Fig. 5.3.

The FSSW was conducted on a two axis FSW machine (Make: ETA Technologies). The sample fixing arrangement on the machine is shown in Fig. 5.4. The FSSW was performed at four different tool plunge depths, 3.2, 3.4, 3.6 and 3.8 mm, keeping other process parameters like tool rotational speed, plunge speed and dwell time constant at 1600 rpm, 8 mm/min and 15 s, respectively. The selection of tool plunge depth is based on several trials made from 3.1 to 3.9 mm in the interval of 0.1 mm. For sandwich sheet at 3.1 mm plunge depth, the joint could not be made due to inadequate plunging and at 3.9 mm plunge depth resulted in excessive thinning of upper sheet at the joint region. The value of tool rotational speed, plunge speed and dwell time are selected based on the available literature on bimetallic FSSW. However, initial FSSW trials for bimetallic sheets show that the joining is possible even at much lower values of welding parameters. So for comparison with sandwich sheets, two extreme ends of plunge depth is selected for bimetallic sheets. The plunge depth is set at 3.2 mm for lower level and 3.8 mm for higher level. All other parameters are kept constant at the same levels (rotational speed: 1600 rpm, plunge speed: 8 mm/min and dwell time: 15 s).

To evaluate the mechanical performance and microscopic analyses of the FSSW joint, five types of sandwich and bimetallic specimens including samples for lap

Table 5.1 Mechanical properties of AA5052-H32 and HDPE sheets

Material	Yield strength (MPa)	Ultimate tensile strength (MPa)	Uniform elongation (%)	Total elongation (%)	Strain hardening coefficient	Strength coefficient (MPa)	Plastic strain ratio
AA5052-H32	155 ± 1	215 ± 1	7 ± 1	9 ± 1	0.16	356 ± 3	0.62
HDPE	–	29 ± 0.5	17.4 ± 2.4	–	–	–	–

Fig. 5.2 Tensile behaviour
of AA5052-H32 and HDPE
sheets

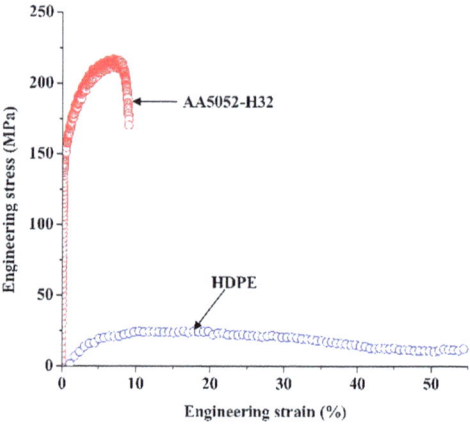

Table 5.2 Chemical
composition of AA5052-H32
(wt%)

Mg	Cr	Zn	Mn	Fe	Si	Cu	Al
3.2	0.23	0.12	0.08	0.20	0.05	0.03	Bal.

Fig. 5.3 FSSW tool used for
the experiments

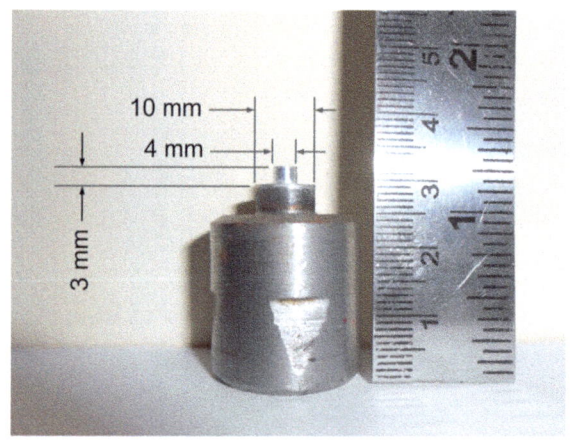

shear test, cross-tension test, peel test, uni-axial tensile test and macro/
microstructural analyses are prepared. AWS B4.0:2007 standard is followed to
prepare the specimen for lap shear test, cross-tension test and peel test. The specimen for uni-axial tensile test and microstructural analyses are non-standard type.
The geometrical dimensions of each type of specimen are schematically shown in
Fig. 5.5. The dimensions remain same for bimetallic sheets as well. The aluminium
and HDPE sheets are cleaned with acetone and soap respectively followed by
drying before making FSSW joint. After cleaning, the three layers are put together
keeping polymer sheet at appropriate position in between two metallic sheets.
A centre point is marked on the upper surface of the sandwich, where the joint has

Fig. 5.4 Fixing
arrangements of specimens
for FSSW

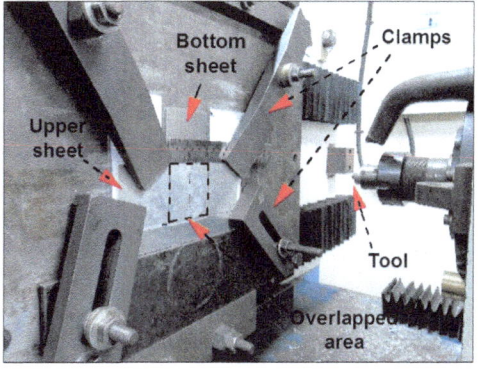

to be made. This point matches with the geometric centre of the overlapped region in the sandwich structure. The sandwich sheet is properly clamped on the bed of the machine and the tool is moved to the centre mark. This tool position is set as the reference zero on the machine. All the predefined process parameters are set to complete the joining operation. To check the repeatability, three specimens are prepared at each set of parameters. Similar procedure is repeated for bimetallic sheets at two extreme set of parameters.

5.2.2 FSSW Joint Characterization

For macro- and microstructural analysis, the FSSW joints are sectioned at the centre using fine hand saw. The half part of the joint is mounted in epoxy resin and polished well using various emery papers varying from coarse to fine grades followed by cloth polishing. The polished sample is then taken for ultrasonic cleaning. Perfectly cleaned surface is chemically etched for 60–90 s in a solution containing HF, HCl, HNO_3, H_2O in the proportion of 2:3:5:10. The etched surface is examined in an optical microscope (Make: Carl Zeiss). Macro features such as hook formation are observed by stitching many photographs taken at 50× magnification. For microstructure, the polished surface is dip etched into the same etchant for 20–22 s. Images are taken at 500× magnification at base zone and a distance 3 mm from keyhole centre or 1 mm from keyhole periphery. Average grain diameter is also measured to perform comparative analysis between sandwich and bimetallic FSSW and to study the effect of plunge depth.

Further, the hardness variation at the joint location is also observed using a micro-hardness tester (Make: Omni Tech). A load of 500 gf and dwell time of 10 s is applied for indentation. The Vickers micro-hardness is measured across the width of the joint at two levels of thickness, mid-thickness of upper sheet and mid-thickness of lower sheet (Fig. 5.6). The distance between two indentations is kept as 1 mm. The hardness is measured on both sides of the joint to check weld symmetricity.

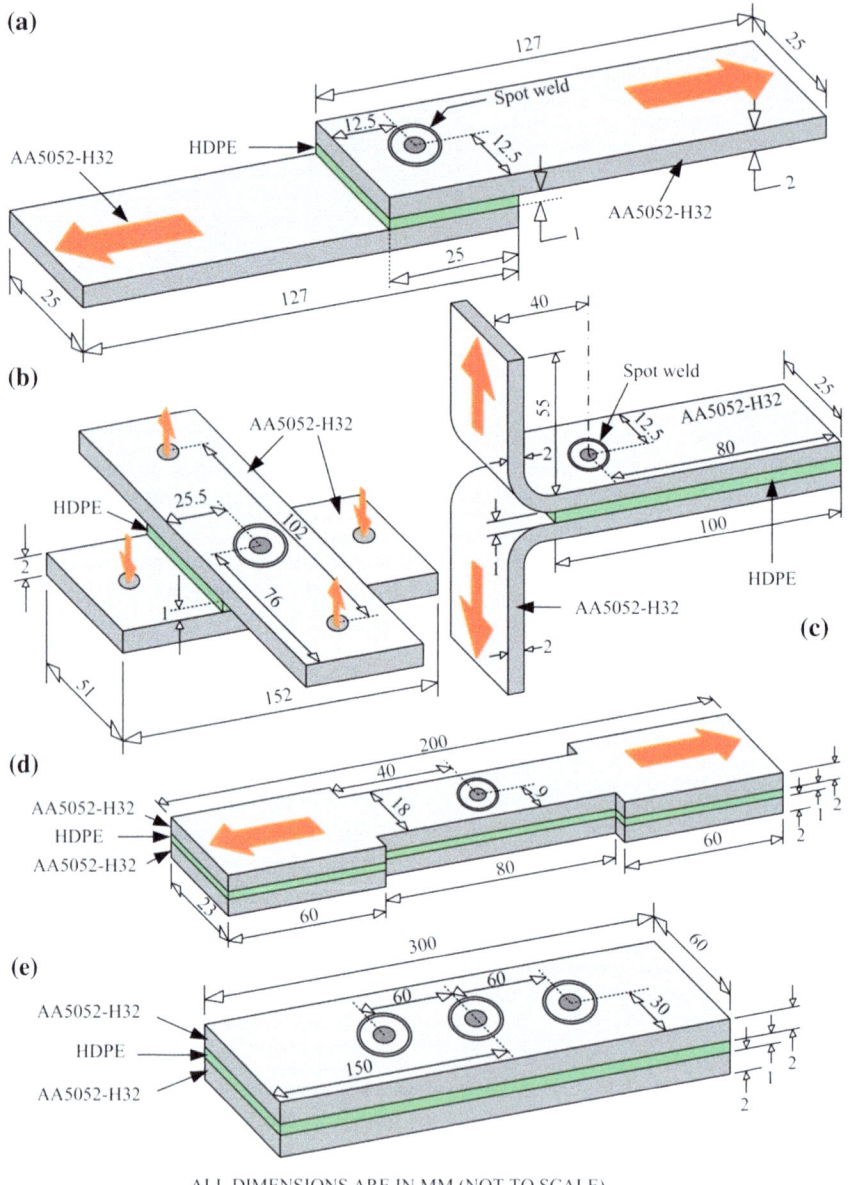

ALL DIMENSIONS ARE IN MM (NOT TO SCALE)

Fig. 5.5 Schematic drawing of specimens for **a** lap shear test, **b** cross-tension test, **c** peel test, **d** uni-axial tensile test, and **e** microstructural analysis

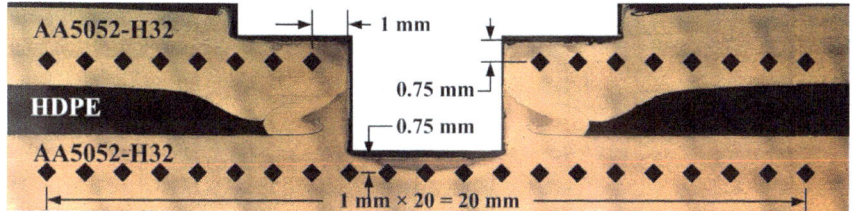

Fig. 5.6 Schematic view of micro-hardness measuring positions in the joint cross-section

To conduct mechanical tests some post-welding minor modifications are done on peel test specimens. The peel test samples are bent in T-shape after welding since it is difficult to weld the bent sheets. The mechanical tests (lap shear, cross-tension and peel) are done on a hydraulically controlled universal testing machine (Model: UTE 20, Make: FIE). The uni-axial tensile test is conducted on a servo-hydraulically driven universal testing machine (Model: UT-04-0250, Make: BISS). All the mechanical tests are done at room temperature and at a ram speed of 1 mm/min.

5.2.3 Temperature Measurement

The distribution of temperature is evaluated in the joint cross-section to investigate the behaviour of sandwich FSSW in presence of polymer. For this, the spot welding is conducted at the edges of the sheets. To measure the temperature, K-type thermocouples are used. Six numbers of thermocouple wires are used; three are attached on upper sheet and three on lower sheet. The placement of thermocouples and physical appearance of the joint at the edge is shown in Fig. 5.7. All the thermocouples are calibrated before using it. Temperature measurement is done on only one side of the joint assuming symmetric weld.

The welding is done at the edge of sample to evaluate temperature distribution across the joint cross-section. In this process the half portion of the pin remains out of the specimen. The problem with FSSW at edge is the continuous expel of stirred

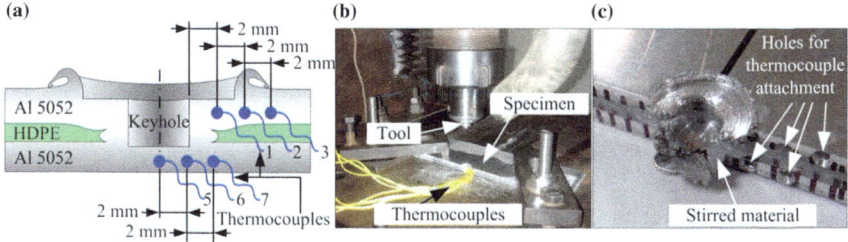

Fig. 5.7 Thermocouple positioning: **a** schematic representation, **b** attachment, and **c** produced joint

Table 5.3 FSSW parameters for comparing temperature measurement

Tool plunge depth (mm)	Tool rotational speed (rpm)	Plunge speed (mm/min)	Dwell time (s)
3.2	1100	4	5
3.8	1500	10	20

material out of the joint, but it always occurs from only one side of the joint. Hence, the thermocouples are attached on the other side of the joint. To attach the thermocouple wires with the specimen, circular holes of 1 mm diameter and 2 mm depth are made at the edge of upper and lower sheets at required position. The tip of the thermocouples are inserted tightly into the holes. Once all thermocouples are attached, the sample is mounted on the machine. The thermocouple wires are also pasted to the bed of the machine in order to avoid detachment during welding. Proper clamping is done, predefined parameters are set and the machine is started. The process is applied for sandwich and bimetallic sheets at 3.2 and 3.8 mm plunge depth. During the process, it is observed that the temperature at each location is continuously varying with time. Therefore, the peak temperature attained at each location till the process completion is selected for analysis. Moreover, the peak temperature developed in actual FSSW is expected to be higher than the FSSW at edge.

The temperature measurement at the edge is also done by infrared (IR) camera. The temperature obtained with this method is compared with that obtained with thermocouple in the same locations. FSSW is done for sandwich and bimetallic sheets at two extreme set of welding parameters as shown in Table 5.3 for comparison.

The upper sheet of the sandwich is selected for the observation and the temperature is measured at the locations identical to that in thermocouple. The infrared camera is kept near the edges of the specimen, which is perpendicular to the tool axis. The image capturing is initiated from start to the end of the process. The emissivity of the material plays a major role to find out the temperature field around the joint. For aluminium and HDPE the emissivity value is selected as 0.35 [22] and 0.95 [8] respectively.

5.3 Results and Discussion

In this section, a detailed examination of FSSW joints of sandwich and bimetallic sheets are done at different tool plunge depths. The FSSW joint characterization is done by evaluating the hook morphologies, flash formation, temperature measurement, microstructural analysis and hardness measurement. Further, the cumulative effect of all these indexes is described through mechanical performance and failure modes.

5.3.1 Hook Formation

The formation of hook like geometry at the interface of two sheets is an important phenomenon during FSSW. The dimensions and shape of the hook are crucial factors which decide the joint behaviour under external loading [34]. In the present study hook formation is observed. The hook always originates from the interface and it forms because of upward movement of trapped oxide layer at the faying surface due to tool plunging [2]. It should be noted here that there are two interfaces in the sandwich sheet; the upper interface and the lower interface. The upper interface represents the region between upper sheet and core, while the lower interface represents the region between core and lower sheet. On the other hand, the bimetallic sheet has only one interface which is between the upper and the lower sheet. Therefore, it is believed that the number of hooks formed will be different for sandwich and bimetallic sheets. Interestingly, not only the number of hooks, but also their characteristics are different in sandwich and bimetallic sheets. A comparison between hook formation in FSSW of sandwich and bimetallic sheet is done and shown in Fig. 5.8. Only one hook is prominently seen for bimetallic sheets though two hook formation is initiated at the interface. These are named as primary and secondary hooks. When the pin plunges into the lower sheet, upward movement of the interface near the joint occur due to backward extrusion of the stirred material beneath pin face. This results in the formation of primary hook. After a while, when the tool shoulder starts plunging the upper sheet, a larger area gets plasticised on the upper sheet. This plasticised material is pushed into the weld by the shoulder. So the upward movement of the material extruded by the pin is restricted by the material coming from the upper sheet. This results in outward flow of stirred material. But this movement is restricted by non-plasticised materials

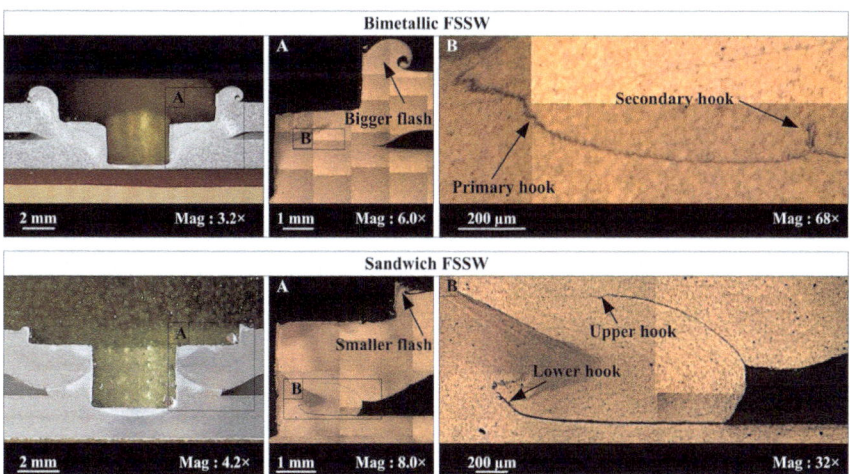

Fig. 5.8 Formation of hook in bimetallic and sandwich sheets during FSSW

surrounding the weld. So, a boundary between plasticised and non-plasticised materials forms above the interface. This results in the formation of a secondary hook. Usually bimetallic FSSW exhibits formation of only one hook, but there exists a tendency to form secondary hook. Solanki et al. [31] have found formation of two distinct hooks in FSSW of magnesium alloy. Rao et al. [25] have reported that primary and secondary may merge together to appear like a single hook depending upon process parameters.

In most of the literature of bimetallic FSSW, formation of single hook is discussed. Badarinarayan et al. [2] have done FSSW on aluminium 5754-O sheets and the micrograph of the weld shows the formation of two hooks, but the secondary hook is so small that effect due to this is insignificant and finally single (primary) hook is analysed. Sometimes, the secondary hook does not form a hook like shape. So it is accepted as a small crack. Rao et al. [27] have found formation of only one hook in FSSW of dissimilar materials.

For sandwich sheet, in the present work, it is observed that two hooks formed; one at the upper interface and another at the lower interface. For analysis the hook formed at upper and lower interface of the sandwich sheet are named as upper hook and lower hook respectively. Similar finding was also observed by Rana et al. [23] for FSSW of sandwich sheets. Unlike bimetallic sheet, the material flow pattern is different in sandwich sheet. Due to presence of polymer in the core, there is an initial gap maintained between upper and lower sheets. When the tool pin starts plunging the lower sheet, the shoulder also starts plunging the upper sheet. So the lower and upper sheet simultaneously getting plasticised. This creates downward flow of plasticised material from upper sheet and upward flow from lower sheet. Both materials find an easy passage to flow into soft polymeric core at the mid-thickness and a large portion of the stirred material is extruded into the core. The oxide layer present at the lower interface gets fragmented and moves upward which leads to the formation of lower hook. It should be noted here that the height of the lower hook in sandwich sheet is lesser than in bimetallic sheet. This happens because the upward flow of lower sheet material quickly gets diverted into the core which ceases upward bending of lower hook. Further, the materials easily flow into the core, so no secondary hook forms at the lower interface, unlike in bimetallic sheet. At the upper interface, it is seen that the hook formation originates from a location much lower than the upper interface because a portion of interface gets suppressed against the polymer core. This happens due to early plasticisation of upper sheet material. Similar to lower interface, no sign of secondary hook is observed at upper interface.

The variation in upper and lower hook geometries with respect to four tool plunge depths is observed for sandwich sheets. The macroscopic view of the joint cross-section of sandwich sheets after FSSW is shown in Fig. 5.9. For bimetallic sheets, joint features are studied at two extreme set of parameters where the tool plunge depth are 3.2 and 3.8 mm at lower and upper extreme respectively. To investigate the effect of tool plunge depth, five geometrical indexes are selected. These are upper bond width, lower bond width, aspect ratio of hook, and effective upper sheet thickness. The upper bond width and lower bond width are the shortest

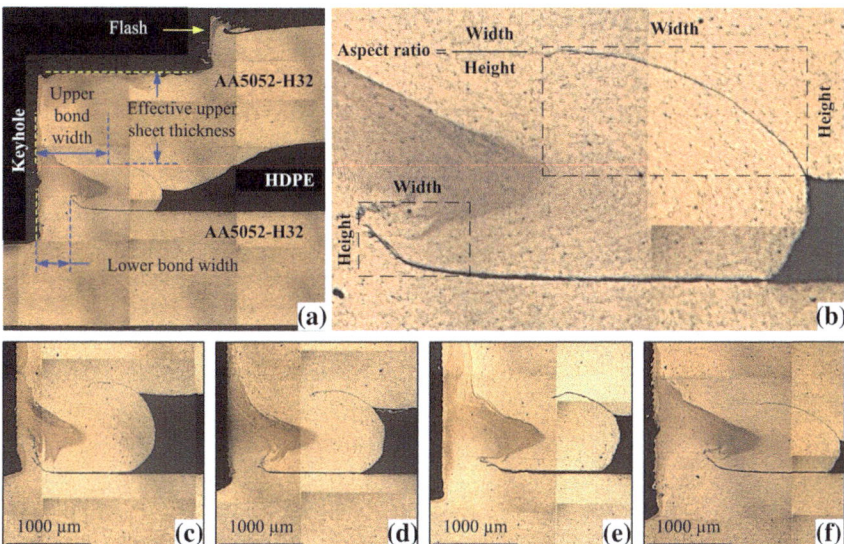

Fig. 5.9 Hook: **a** geometrical nomenclatures, **b** aspect ratio, and formation at tool plunge depth of **c** 3.2 mm, **d** 3.4 mm, **e** 3.6 mm, **f** 3.8 mm

distance between keyhole periphery and tip of the upper and lower hook. Aspect ratio is the ratio of hook width to hook height. The effective upper sheet thickness is the shortest distance between upper hook and the surface on which the tool shoulder touches. It should be noted here that the nomenclatures associated with the secondary and primary hooks in bimetallic sheet are similar to that with upper and lower hook in sandwich sheets.

The geometrical measurements of various joint features in sandwich and bimetallic sheets are listed in Table 5.4. It is found that in both the systems, at all plunge depths, the upper bond width is greater than lower bond width. This is attributed to greater heat generation underneath the shoulder surface than at the probe surface due to larger contact area at shoulder surface. Increased heat generation is responsible for higher peak temperature which promotes plasticisation of material. As the plasticisation increases, the outward flow of material also increases resulting in larger bonded width in the vicinity of upper hook. Further, with increasing tool plunge depth, the upper and lower bond width increases for sandwich and bimetallic sheets. When plunge depth increases, more material displacement occurs in the joint region to accommodate the tool. A part of the increased amount of displaced material is contributed by the expelled flash and remaining is trapped within. When plasticised material in the joint region tries to accommodate, it flows away from the keyhole, thereby enlarging bonded width.

Apart from the outward material flow, many other phenomena also affect the bonded width during change of plunge depth. Tutar et al. [33] have also reported that increase in plunge depth naturally increase the heat exposure time which in turn

Table 5.4 FSSW weld geometrical measurements of the joint morphologies

System[a]	Plunge depth (mm)	Upper bond width (mm)	Lower bond width (mm)	Aspect ratio[b] UH (or) SH	LH (or) PH	Effective upper sheet thickness (mm)
SW	3.2	0.91	0.18	8.23	0.97	1.93
	3.4	1.01	0.46	5.58	2.67	1.84
	3.6	1.28	0.56	4.11	3.21	1.79
	3.8	1.29	0.66	2.36	3.41	1.60
BM	3.2	1.72	0.62	0.53	2.20	1.26
	3.8	2.23	1.28	0.09	4.36	0.91

[a]*SW* sandwich sheets; *BM* bimetallic sheets
[b]*UH* upper hook; *SH* secondary hook; *LH* lower hook; *PH* primary hook; *UH and LH* are for sandwich sheets; *SH and PH* are for bimetallic sheets

widens the weld zones. Further, bimetallic sheets show larger bonded width in comparison to sandwich sheet. This is likely due to larger outward flow of material in bimetallic sheet in absence of polymeric core.

Another feature is the aspect ratio of hook which is the ratio of hook width to hook height. It is observed that with increasing tool plunge depth, aspect ratio of upper hook in sandwich sheet decreases. Decrease in aspect ratio means decreasing width and increasing height. This is likely due to the lowering of hook origin because of increased downward material flow from shoulder region. In bimetallic sheet, since there is no easy passage in the middle for increased material flow from upper sheet, the downward flow of material diverted into outward flow restricts the growth of secondary hook. This results in the formation of steeper hook having larger height and lesser width. The secondary hook aspect ratio in bimetallic sheet is much lesser than upper hook ratio in sandwich sheet. However, completely reverse trend is observed for lower hook and primary hook in sandwich and bimetallic sheet respectively. In sandwich sheet, width of lower hook remains almost constant, but hook height decreases with increasing tool plunge depth because the upward growth of lower hook is annihilated by increased material flow into the core. This results in increasing lower hook aspect ratio. On the other hand, in bimetallic sheet, the vertical growth of primary hook is promoted at higher plunge depth since a large portion of probe goes into the lower sheet.

The effective upper sheet thickness decreases with increasing tool plunge depth. This is due to upper sheet thinning with increase in plunge depth. In comparison, the sandwich sheet shows larger effective upper sheet thickness than bimetallic sheet. Literature suggests that larger effective upper sheet thickness gives better joint strength in FSSW [26].

Another observation is flash formation. It depends upon the displaced material due to tool plunging. In sandwich FSSW, when tool plunging starts, the shoulder plunges into a part of the upper sheet and the pin plunges into a part of the lower sheet, while the pin is fully penetrated through the polymer core. So the displaced volume of material is the contribution of upper and lower sheet. A little contribution from the polymer can be ignored due to lower density. In bimetallic FSSW, the contribution is totally from upper and lower sheet. At any tool plunge depth level, the pin plunging into lower sheet is not same for sandwich and bimetallic sheet due to difference in their total thickness, but the shoulder plunge depth will remain same. The pin plunge depth in bimetallic sheet will always be 1 mm greater than that in sandwich sheet as shown in Fig. 5.10. Hence, the amount of material displaced is also different in both systems.

The ideal value of material displaced can be calculated as the volume of tool inside the system. At different levels of tool plunge depth, ideal volume displaced is calculated and shown in Table 5.5. It is seen that material displacement in sandwich sheet is always lesser than bimetallic sheet at any particular plunge depth.

The expelled flash volume is calculated as below. The contour of expelled flash is mapped using AutoCad software. The closed contour is revolved by $360°$ about a circle of diameter equal to shoulder diameter of the tool (10 mm). This gives a virtual solid geometry almost equal to flash formed in actual case. The volume of the virtual flash is calculated in the software.

A comparison is done between sandwich and bimetallic sheets at two extreme set of parameters. The geometrical measurement of flash is given in Fig. 5.11. In sandwich sheets, lesser amount of flash is expelled from the system than the bimetallic sheets. For sandwich sheet, the ideal volume of material displaced is also lesser and a large portion of stirred material gets accommodated in the soft polymeric core. The ideal displaced volume in sandwich sheet at 3.8 mm tool plunge depth is about 88 mm^3, while it is about 100 mm^3 in bimetallic sheet at the same plunge depth (Table 5.5). Hence in this case the flash volume of sandwich sheet is much smaller than in bimetallic sheet—3.46 mm^3 versus 57.63 mm^3 (Fig. 5.11). It can be concluded that displaced material is accommodated inside the core in sandwich sheet. Flash is a defect and hence lesser flash formation is always good. Moreover, it does not contribute in load bearing. The flash volume is higher at larger tool plunge depth due to increased material displacement. This has been substantiated by Oladimeji et al. [18] through expelled flash volume calculations in the FSSW of aluminium alloys.

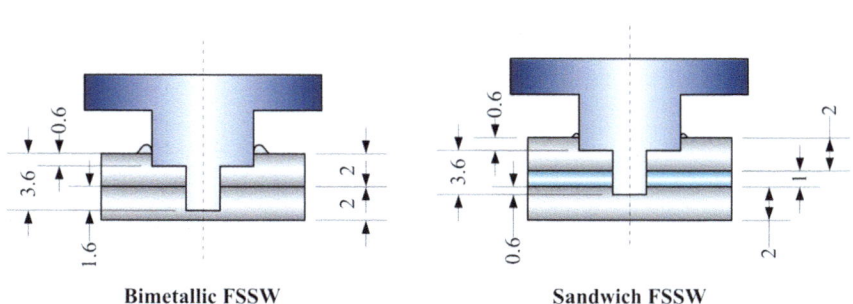

Bimetallic FSSW   **Sandwich FSSW**

Fig. 5.10 Difference in pin plunge depth in bimetallic and sandwich FSSW

Table 5.5 Ideal volume of displaced material during FSSW

System	Tool plunge depth[a] (mm)	Tool shoulder				Tool pin				Total displaced volume (mm³)
		Diameter (mm)	Surface area (mm²)	Plunge depth (mm)	Vol. (mm³)	Diameter (mm)	Surface area (mm²)	Height (mm)	Vol. (mm³)	
SW	3.2	10	78.54	0.2	15.71	4	12.57	3	25.13	40.84
	3.4			0.4	31.42					56.55
	3.6			0.6	47.12					72.26
	3.8			0.8	62.83					87.96
BM	3.2			0.2	15.71				37.70	53.41
	3.8			0.8	62.83					100.53

[a]Tool plunge depth = shoulder plunge depth + pin height

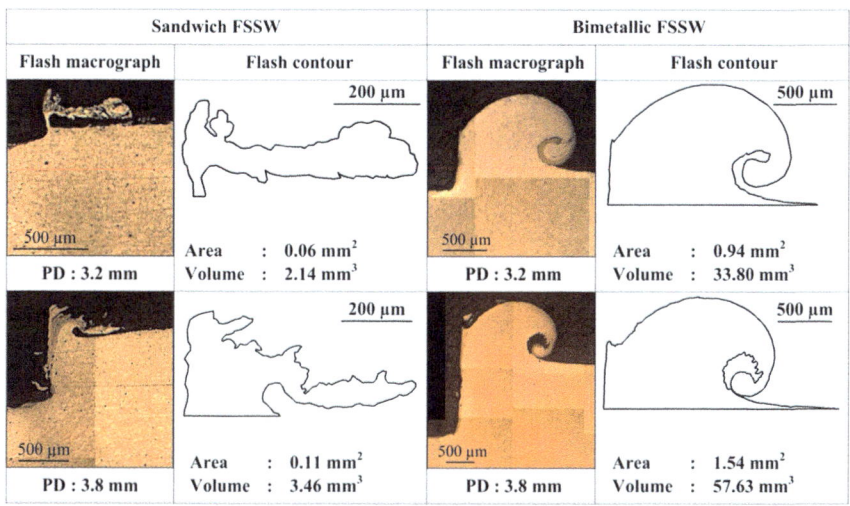

Fig. 5.11 Volume change in flash for sandwich and bimetallic sheet

5.3.2   Temperature Evaluation

The distribution of maximum temperature attained around the joint during FSSW is shown in Fig. 5.12. Higher temperature field is observed in the keyhole vicinity (Fig. 5.12a). The peak temperature distribution by IR camera and thermocouple is compared in the upper sheet of sandwich as shown in Fig. 5.12b. Preliminary observation suggest that the temperature recorded by the thermocouple is lesser than the IR camera. This is due to the presence of holes at the edges, which reduces the frictional force between tool and sheet. Also, the accurate value of material emissivity is difficult to find. The temperature data obtained by thermocouple is found more reliable, so selected for further analysis. Higher peak temperature is observed at larger plunge depth. This is attributed to increased heat input associated to larger material deformation, larger tool-work piece interface area and increased processing time. Further, the influence of plunge depth in sandwich sheet is smaller than bimetallic sheet. This is confirmed by lesser temperature difference in sandwich sheet ($\Delta T_{SW} = 129$ °C) than bimetallic sheet ($\Delta T_{BM} = 218$ °C) near the keyhole as shown in Fig. 5.12c. Similar observation is found in lower sheet (Fig. 5.12d) also ($\Delta T_{SW} = 9$ °C, $\Delta T_{BM} = 114$ °C).

In comparison to bimetallic sheet, lesser peak temperature is recorded in sandwich sheet. It likely is due to lesser frictional contact area between tool and the metallic sheet in sandwich. The temperature gradient through thickness is lesser for sandwich sheet especially at higher plunge depth. The temperature gradient through thickness is measured by the difference in peak temperature of upper and lower sheet at a distance 4 mm from keyhole centre. Further, the temperature gradient with respect to the distance from keyhole in upper sheet of sandwich sheet is lesser

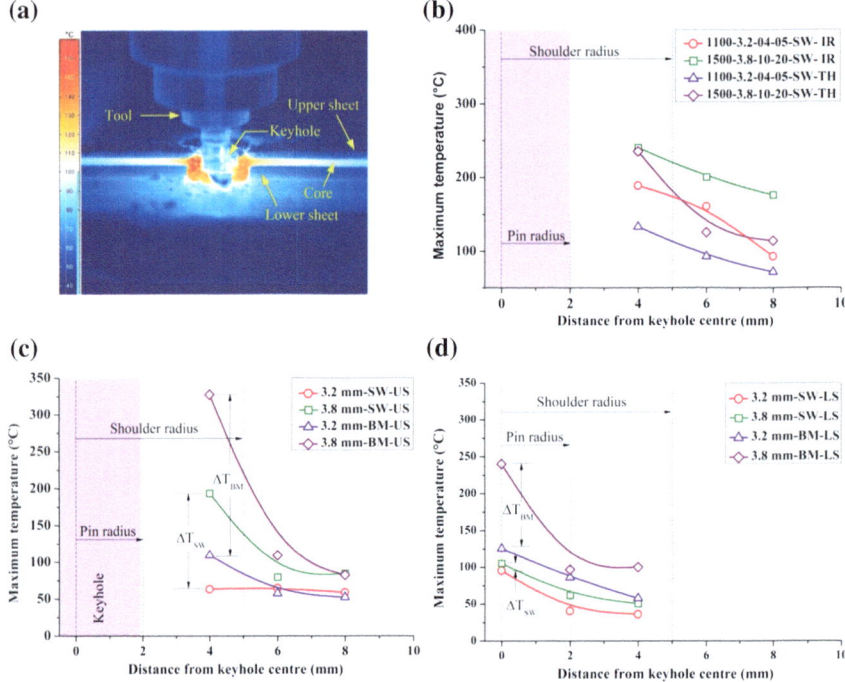

Fig. 5.12 a Infrared image of FSSW at edge; peak temperature distribution at mid-thickness of **b** the upper sheet of SW in both measuring techniques, **c** the upper sheets of SW and BM measured by thermocouples, and **d** the lower sheets of SW and BM measured by thermocouples [SW: sandwich sheets; BM: bimetallic sheets; US: upper sheet; LS: lower sheet]; the sequence in the legend of (**b**) is tool rotational speed-plunge depth-plunge speed-dwell time-measuring method

than that in bimetallic sheet. In contrast, the temperature of upper sheet is larger than the lower sheet. This confirms the fact that major part of the heat generated in FSSW comes from friction between tool shoulder and work-piece. Moreover, a part of heat generated at the lower sheet gets transferred into the base plate. Further, the peak temperature gradient with respect to distance from keyhole in the upper sheet is larger than in lower sheet.

5.3.3 Grain Size and Hardness Distribution

The microstructure of the upper sheet on the joint at 1 mm distance from the keyhole periphery and parent metal are shown in Fig. 5.13a–e. The measured grain size is plotted in Fig. 5.13f. With respect to parent metal microstructure, grain refinement occurred due to dynamic recrystallization at high temperature in all the FSSW joints. However, Fig. 5.13a for sandwich sheet at 3.2 mm plunge depth

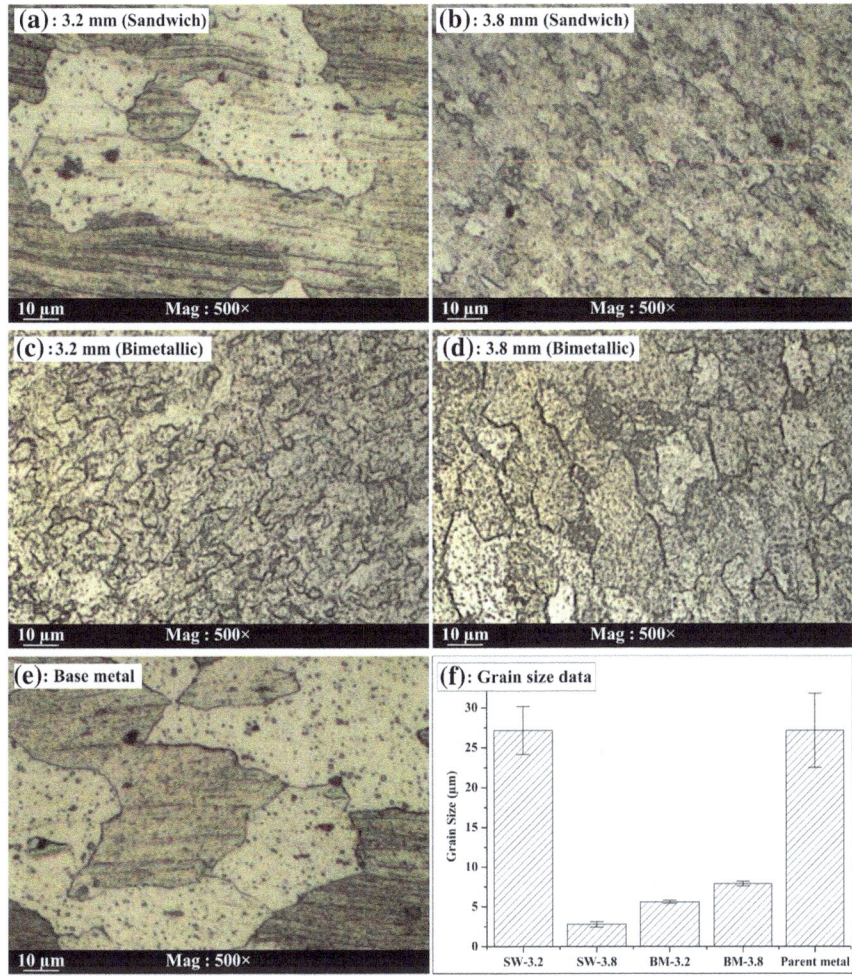

Fig. 5.13 Microstructure of the upper sheet at a distance of 1 mm from keyhole periphery of **a** 3.2 mm-sandwich, **b** 3.8 mm-sandwich, **c** 3.2 mm-bimetallic, **d** 3.8 mm-bimetallic, **e** base metal microstructure, and **f** grain size measurement [SW: sandwich sheets; BM: bimetallic sheets]

reveal almost same grain size with respect to parent metal. This is due to the lowest peak temperature (refer Fig. 5.12c) restricting the dynamic recrystallization to occur within the identified zone for grain size evaluation.

With increasing plunge depth, grain size increases in bimetallic sheets (Fig. 5.13c, d, f). This is attributed to grain growth at higher temperature at higher plunge depth as seen in Fig. 5.12c on temperature distribution. In contrast, at 3.8 mm plunge depth, the sandwich sheet shows finer grains as compared to 3.2 mm plunge depth, though temperature developed is larger in 3.8 mm plunge depth (Fig. 5.12c). This is probably because at 3.2 mm plunge depth, the effect is insignificant to change the grain size in case of sandwich. In this case, the deformation provided and temperature rise are insufficient and hence grain size remains almost same as that of parent sheet. When the plunge depth is increased to 3.8 mm, though there exist deformation and temperature rise, it is insufficient to increase the grain size to the extent seen in 3.2 mm plunge depth (Fig. 5.13f). There is a large difference in grain size between sandwich and bimetallic sheet only at lower plunge depth, but not at higher plunge depth. This is attributed mainly due to insignificant change in grain size of sandwich sheet at 3.2 mm as compared to parent sheet (Fig. 5.13a, e, f).

The variation in hardness across the joint cross-section is evaluated and shown in Fig. 5.14. It is observed that the hardness is symmetrical about the keyhole axis and maximum hardness is attained in region closer to the tool or keyhole. The reason behind this is the degree of dynamic recrystallization is more in the vicinity of tool

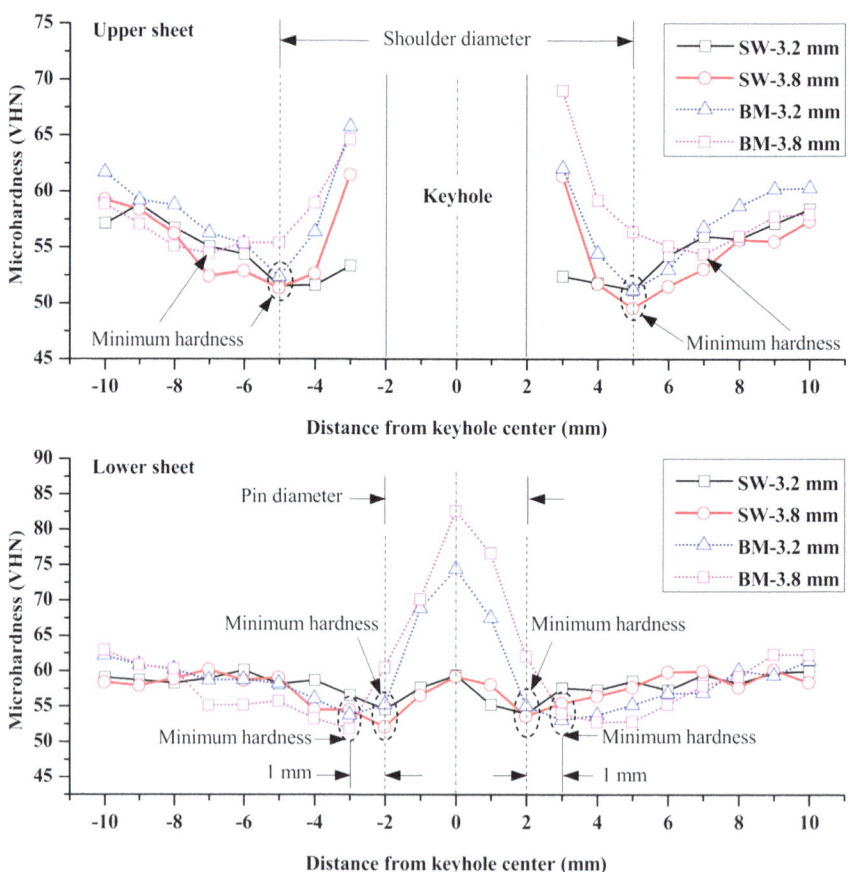

Fig. 5.14 Hardness distribution at mid-thickness of the upper and lower sheet. Error variation: ±1.5 VHN [SW: sandwich sheets; BM: bimetallic sheets]

because of larger heat generation. With increasing tool plunge depth, the hardness increases in sandwich sheet due to grain refinement. In bimetallic sheet, the hardness increases with increasing plunge depth though grain size increases. It is reported by Huskins et al. [13] that strengthening mechanism of AA5XXX alloys are not straightway controlled by grain size, rather it is mainly governed by dislocation density. Sato et al. [30] has reported that distribution of second phase particles also plays a major role in controlling the hardness of AA5XXX alloys. It is believed that at higher plunge depth, intense stirring due to higher temperature would resulted in fragmentation of second phase particles. This could have possibly improved the hardness of bimetallic sheet at higher plunge depth. The bimetallic sheets are slightly harder than the sandwich sheets (Fig. 5.14). This is mainly due to increased stirring. The bimetallic sheet exhibits larger temperature evolution than sandwich sheets in most of the cases (Fig. 5.12), which resulted in increased intensity of stirring.

In upper sheet, hardness varies prominently within shoulder diameter, while it is pin diameter in lower sheet. The lowest hardness location in upper sheet exists farther from keyhole than in lower sheet. The location of minimum hardness in the upper sheet of sandwich system coincides with that in bimetallic sheet and lies just below the shoulder edge. However, in lower sheet the minimum hardness location lies below the probe edge for sandwich sheet, but it is 1 mm away from the probe edge for bimetallic sheet. It is known that in FSSW of AA5XXX alloys, metallurgical recovery causes localised softening [14]. The location of this zone lies where dynamic recrystallization ceases and material undergoes only thermal cycle, typically known as heat-affected zone (HAZ). The farther existence of minimum hardness location in lower sheet of bimetallic system is due to higher temperature which shifts the HAZ away from keyhole.

5.3.4 Mechanical Performance and Failure Modes

The effect of tool plunge depth on the joint strength and extension at failure of sandwich and bimetallic sheets during various loading conditions is shown in Fig. 5.15. Due to the complex geometry of FSSW joint and typical shapes of specimens, the joint strength is characterised by maximum load attained during test and ductility is related with extension up to maximum load.

With increasing tool plunge depth, joint strength and ductility considerably improved in sandwich sheets. In lap shear test (Fig. 5.15a), the joint strength of the sandwich sheet consistently increases up to 3.8 mm plunge depth. The uniform elongation increases up to 3.4 mm plunge depth, then remains same. In cross-tension test and peel tests, a consistent improvement in joint strength and ductility is observed with increasing tool plunge depth. It should be noted here that the extension recorded in cross-tension and peel test is the combined effect of the sheet bending and elongation in the joint. In uni-axial tensile test, influence of tool plunge depth is not much in sandwich sheet. Better joint performance of sandwich

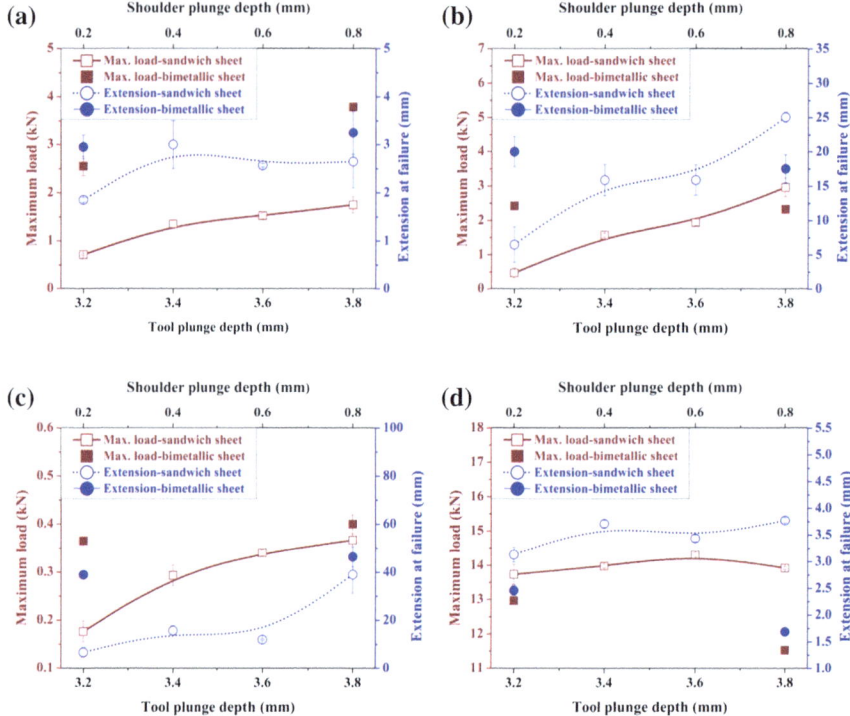

Fig. 5.15 Maximum load and extension at failure for sandwich and bimetallic sheets: **a** lap-shear test, **b** cross-tension test, **c** peel test, and **d** uni-axial tensile test

sheet at higher plunge depth is associated with larger bond width and larger joint hardness. During mechanical testing, crack forms at the weakest section in the joint. It is evident from Fig. 5.14 that the joint region near the lower sheet is weaker than upper sheet. Further, the hooks are the favourable site for crack initiation. So, the failure of the joint begins at the lower hook tip and the crack propagates towards the keyhole. The lower bond width represents the length of the crack propagation path and decides the failure load. Since the lower bond width increases with tool plunge depth, the failure load will also increase. Larger value of upper bond width and joint hardness at higher plunge depth in upper sheet further adds some strength into the joint. However, the effect of other joint features such as hook aspect ratio and effective upper sheet thickness are insignificant although they are affected by process parameters. The effect of tool plunge depth on the joint behaviour of bimetallic sheet is found to be significant. In lap shear test and peel test, the joint performance is improved, while it is deteriorated in cross-tension test and uni-axial tensile test.

A comparative analysis of failure load of sandwich and bimetallic system for each test at two plunge depths, 3.2 and 3.8 mm, is performed in Table 5.6. The improvement in joint strength of sandwich sheet in all loading conditions is much

Table 5.6 Comparative assessment of joint strength with respect to plunge depth

System/ PD (mm)		Lap shear test		Cross-tension test		Peel test		Uni-axial tensile test	
		F_L (kN)	ΔF_L (%)	F_C (kN)	ΔF_C (%)	F_P (kN)	ΔF_P (%)	F_U (kN)	ΔF_U (%)
SW	3.2	0.67	114.85	0.47	530.50	0.18	107.55	13.74	1.37
	3.8	1.45		2.96		0.37		13.93	
BM	3.2	2.55	48.63	2.42	−3.72	0.37	8.11	12.97	−11.10
	3.8	3.79		2.33		0.40		11.53	
3.2	SW	0.67	280.60	0.47	414.89	0.18	105.56	13.74	−5.60
	BM	2.55		2.42		0.37		12.97	
3.8	SW	1.45	161.38	2.96	−21.28	0.37	8.11	13.93	−17.23
	BM	3.79		2.33		0.40		11.53	

higher than the bimetallic sheet. The maximum influence of tool plunge depth for sandwich system is seen in cross-tension test, while it is lap shear test for bimetallic system. This particular analysis suggests that the sandwich FSSW can be greatly employed in situation when the sandwich is subjected to loading along the geometrical axis of the joint.

The percentage difference in failure load of bimetallic sheet with respect to sandwich sheet is calculated at 3.2 and 3.8 mm (Table 5.6) plunge depths individually for each test. The bimetallic system produces superior joint at lower plunge depth than at higher plunge depth. Further, at 3.2 mm plunge depth, best joint performance of bimetallic sheet in comparison to sandwich sheet is observed in cross-tension test, while better joint performance of sandwich sheet is observed in uni-axial tensile test. Moreover, better joint performance shown by sandwich sheet in uni-axial tensile test at 3.2 mm plunge depth. At 3.8 mm plunge depth, difference in failure loads of both systems is minimised. Also, the joint performance of sandwich sheet is superior in cross-tension and uni-axial tensile test. From the comparison shown in Fig. 5.15 and Table 5.6, it is concluded that the sandwich sheet gives optimum joint property at tool plunge depth of 3.6 mm and beyond.

The physical appearance of the fracture surfaces of tested sandwich sheet is shown in Fig. 5.16. It is observed that mode of fracture is 'nugget pullout' for all specimens at any plunge depth in lap shear, cross-tension and peel test. The nugget pullout mode of failure is attributed to the smaller value of lower bond width when compared to upper bond width. This is observed in Fig. 5.9 and Table 5.4. The size of upper and lower bond width depends upon the position of upper and lower hook respectively. These hooks are the geometrical defect in FSSW which always reduces the joint integrity and are favourable site for crack propagation. During loading, crack propagates through upper and lower hooks. Early reach of crack through lower hook at the keyhole is expected since the lower hook tip is nearer to the keyhole. This phenomenon creates a situation of lower sheet detachment from the bond. So the nugget or weld remains attached to the upper sheet in the 'nugget pullout' fracture.

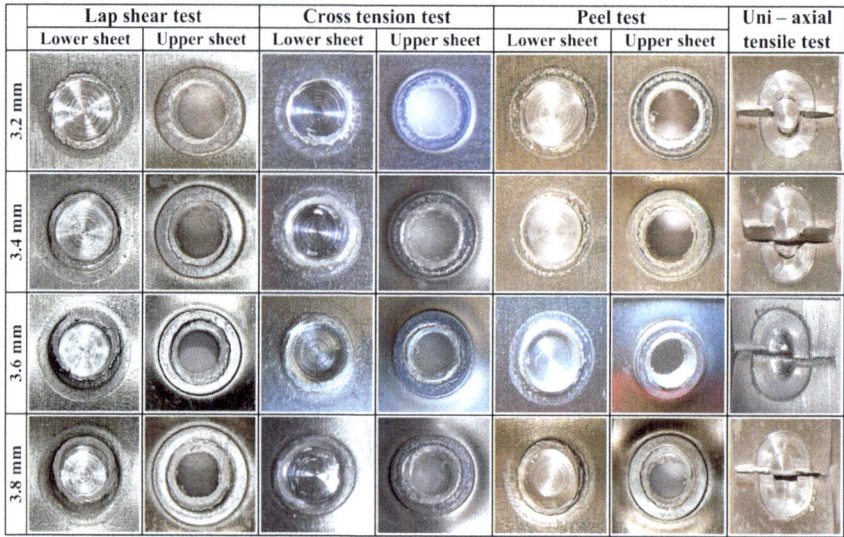

	Lap shear test		Cross tension test		Peel test		Uni – axial
	Lower sheet	Upper sheet	Lower sheet	Upper sheet	Lower sheet	Upper sheet	tensile test
3.2 mm							
3.4 mm							
3.6 mm							
3.8 mm							

Fig. 5.16 Failure modes of sandwich sheets

Table 5.7 Failure modes of bimetallic sheets

Tool plunge depth (mm)	Lap shear test	Cross-tension test	Peel test
3.2	Interfacial shear	Interfacial separation	Interfacial separation
3.8	Interfacial shear	Nugget pullout	Partial sheet fracture

The failure mode is completely different in uni-axial tensile test due to different loading conditions. The crack initiates at the keyhole boundary and starts propagating outward till complete fracture of upper sheet. Once the upper sheet fractures, entire load is carried by lower sheet, which fails almost like in standard uni-axial tensile tests. The failure modes of bimetallic sheet namely nugget fracture and interface separation is shown in Table 5.7.

5.4 Conclusions

In the present work, effect of tool plunge depth on the FSSW of three layered AA5052-H32/HDPE/AA5052-H32 is investigated. The following are the important findings:

1. FSSW joint of sandwich sheet with acceptable performance is obtained at tool plunge depth of 3.6 mm and greater.
2. Two hooks are formed, one at each interface in sandwich sheet. In bimetallic sheet, two hooks are formed at single interface. But one of the hooks (secondary

hook) is much smaller as compared to the other one in bimetallic sheet. However, no direct dependence of joint strength on hook geometry is observed, instead it depends on bond width and joint hardness.

3. The expelled flash volume increases with tool plunge depth in sandwich and bimetallic sheets. However, in comparison to bimetallic sheet, lesser flash forms in sandwich sheet due to larger material volume accommodation in polymeric core.

4. Temperature measurement data reveals that highest temperature is recorded on the upper sheet of bimetallic system at higher plunge depth. This is obtained because larger frictional contact area between tool and metal.

5. The microstructural observation closer to the keyhole of sandwich and bimetallic FSSW reveal finer grains in sandwich sheet due to lesser peak temperature.

6. Deterioration of joint performance is observed in sandwich sheets due to presence of polymer. Formation of smaller bond width at lower sheet side is responsible for this. However, at higher plunge depth superior joint performance of sandwich sheet is observed in cross-tension and uni-axial tensile tests. Though this the case, the flash formation is lesser in sandwich sheets as compared to bimetallic. By using sandwich sheets, the flash defect can be minimised.

7. In all the tests, nugget pullout mode of failure occurred due to smaller value of lower bond width. The failure mode is also found to be independent of the test method. So, it is confirmed that failure mode in the present study depend only on relative size of lower bond width.

Acknowledgements Authors thank CIF, IIT Guwahati for extending the universal testing machine facility for testing the joints. The authors also thank Department of Science and Technology, India for funding a FIST project (ETII-244/2008) by which the infrared camera was procured and used for temperature measurement. The present work is not funded by any funding organisation

References

1. Arici, A., Mert, S.: Friction stir spot welding of polypropylene. J. Reinf. Plast. Compos. **27**, 2001–2004 (2008). https://doi.org/10.1177/0731684408089134
2. Badarinarayan, H., Shi, Y., Li, X., Okamoto, K.: Effect of tool geometry on hook formation and static strength of friction stir spot welded aluminum 5754-O sheets. Int. J. Mach. Tools Manuf. **49**, 814–823 (2009). https://doi.org/10.1016/j.ijmachtools.2009.06.001
3. Barnes, T., Pashby, I.: Joining techniques for aluminium space frames used in automobiles. J. Mater. Process. Technol. **99**, 62–71 (2000). https://doi.org/10.1016/s0924-0136(99)00367-2
4. Bilici, M.K., Yükler, A.I.: Influence of tool geometry and process parameters on macrostructure and static strength in friction stir spot welded polyethylene sheets. Mater. Des. **33**, 145–152 (2012). https://doi.org/10.1016/j.matdes.2011.06.059

5. Bilici, M.K., Yükler, Aİ., Kurtulmuş, M.: The optimization of welding parameters for friction stir spot welding of high density polyethylene sheets. Mater. Des. **32**, 4074–4079 (2011). https://doi.org/10.1016/j.matdes.2011.03.014

6. Burchitz, I., Boesenkool, R., van der Zwaag, S., Tassoul, M.: Highlights of designing with Hylite—a new material concept. Mater. Des. **26**, 271–279 (2005). https://doi.org/10.1016/j.matdes.2004.06.021

7. Filho, S.T.A., dos Santos, J.F.: Joining of polymers and polymer-metal hybrid structures: recent developments and trends. Polym. Eng. Sci. **49**, 1461–1476 (2009)

8. Genna, S., Leone, C., Tagliaferri, V.: Characterization of laser beam transmission through a high density polyethylene (HDPE) plate. Opt. Laser Technol. **88**, 61–67 (2017). https://doi.org/10.1016/j.optlastec.2016.08.010

9. Gerlich, A., Avramovic-Cingara, G., North, T.H.: Stir zone microstructure and strain rate during Al 7075-T6 friction stir spot welding. Metall. Mater. Trans. A **37**, 2773–2786 (2006). https://doi.org/10.1007/bf02586110

10. Gibson, L.J., Ashby, M.F.: Cellular Solids: Structure and Properties, p. 345. Cambridge University Press (1997)

11. Gower, H.L., Pieters, R.R.G.M., Richardson, I.M.: Pulsed laser welding of metal-polymer sandwich materials using pulse shaping. J. Laser Appl. **18**, 35–41 (2006). https://doi.org/10.2351/1.2080307

12. Hao, M., Osman, K.A., Boomer, D.R., Newton, C.J.: Developments in characterization of resistance spot welding of aluminum. Weld. J. **75**, 1–s–8–s (1996)

13. Huskins, E.L., Cao, B., Ramesh, K.T.: Strengthening mechanisms in an Al-Mg alloy. Mater. Sci. Eng. A **527**, 1292–1298 (2010). https://doi.org/10.1016/j.msea.2009.11.056

14. Kesharwani, R.K., Panda, S.K., Pal, S.K.: Experimental investigations on formability of aluminum tailor friction stir welded blanks in deep drawing process. J. Mater. Eng. Perform. **24**, 1038–1049 (2014). https://doi.org/10.1007/s11665-014-1361-5

15. Kim, K.J., Kim, D., Choi, S.H., Chung, K., Shin, K.S., Barlat, F., Oh, K.H., Youn, J.R.: Formability of AA5182/polypropylene/AA5182 sandwich sheets. J. Mater. Process. Technol. **139**, 1–7 (2003). https://doi.org/10.1016/s0924-0136(03)00173-0

16. Lee, C.-Y., Choi, D.-H., Yeon, Y.-M., Jung, S.-B.: Dissimilar friction stir spot welding of low carbon steel and Al–Mg alloy by formation of IMCs. Sci. Technol. Weld. Join. **14**, 216–220 (2009). https://doi.org/10.1179/136217109x400439

17. Mitlin, D., Radmilovic, V., Pan, T., Chen, J., Feng, Z., Santella, M.L.: Structure-properties relations in spot friction welded (also known as friction stir spot welded) 6111 aluminum. Mater. Sci. Eng. A **441**, 79–96 (2006). https://doi.org/10.1016/j.msea.2006.06.126

18. Oladimeji, O.O., Taban, E., Kaluc, E.: Understanding the role of welding parameters and tool profile on the morphology and properties of expelled flash of spot welds. Mater. Des. **108**, 518–528 (2016). https://doi.org/10.1016/j.matdes.2016.07.013

19. Oliveira, P.H.F., Amancio-Filho, S.T., Dos Santos, J.F., Hage, E.: Preliminary study on the feasibility of friction spot welding in PMMA. Mater. Lett. **64**, 2098–2101 (2010). https://doi.org/10.1016/j.matlet.2010.06.050

20. Paidar, M., Sadeghi, F., Najafi, H., Khodabandeh, A.R.: Effect of pin and shoulder geometry on stir zone and mechanical properties of friction stir spot-welded aluminum alloy 2024-T3 sheets. J. Eng. Mater. Technol. Trans. ASME **137**, 3–9 (2015). https://doi.org/10.1115/1.4030197

21. Pickin, C.G., Young, K., Tuersley, I.: Joining of lightweight sandwich sheets to aluminium using self-pierce riveting. Mater. Des. **28**, 2361–2365 (2007). https://doi.org/10.1016/j.matdes.2006.08.003

22. Raikoty, H., Ahmed, I., Talia, G.E.: High speed friction stir welding: a computational and experimental study. Proc. ASME Summer Heat Transf. Conf. **3**, 431–436 (2005). https://doi.org/10.1115/ht2005-72833

23. Rana, P.K., Narayanan, R.G., Kailas, S.V.: Influence of rotational speed on the friction stir spot welding of polymer core sandwich sheets. In: Proceedings of 6th International & 27th All India Manufacturing Technology, Design and Research Conference (AIMTDR-2016). pp. 926–930 (2016)
24. Rana, P.K., Narayanan, R.G., Kailas, S.V.: Effect of rotational speed on friction stir spot welding of AA5052-H32/HDPE/AA5052-H32 sandwich sheets. J. Mater. Process. Technol. **252**, 511–523 (2018)
25. Rao, H.M., Jordon, J.B., Barkey, M.E., Guo, Y.B., Su, X., Badarinarayan, H.: Influence of structural integrity on fatigue behavior of friction stir spot welded AZ31Mg alloy. Mater. Sci. Eng. A **564**, 369–380 (2013). https://doi.org/10.1016/j.msea.2012.11.076
26. Rao, H.M., Rodriguez, R.I., Jordon, J.B., Barkey, M.E., Guo, Y.B., Badarinarayan, H., Yuan, W.: Friction stir spot welding of rare-earth containing ZEK100 magnesium alloy sheets. Mater. Des. **56**, 750–754 (2014). https://doi.org/10.1016/j.matdes.2013.12.034
27. Rao, H.M., Yuan, W., Badarinarayan, H.: Effect of process parameters on mechanical properties of friction stir spot welded magnesium to aluminum alloys. Mater. Des. **66**, 235–245 (2015). https://doi.org/10.1016/j.matdes.2014.10.065
28. Rao, M.D.: Recent applications of viscoelastic damping for noise control in automobiles and commercial airplanes. J. Sound Vib. **262**, 457–474 (2003). https://doi.org/10.1016/s0022-460x(03)00106-8
29. Salonitis, K., Drougas, D., Chryssolouris, G.: Finite element modeling of penetration laser welding of Sandwich materials. Phys. Procedia **5**, 327–335 (2010). https://doi.org/10.1016/j.phpro.2010.08.059
30. Sato, Y.S., Park, S.H.C., Kokawa, H.: Microstructural factors governing hardness in friction-stir welds of solid-solution-hardened Al alloys. Metall. Mater. Trans. A **32**, 3033–3042 (2001). https://doi.org/10.1007/s11661-001-0178-7
31. Solanki, K.N., Jordon, J.B., Whittington, W., Rao, H., Hubbard, C.R.: Structure-property relationships and residual stress quantification of a friction stir spot welded magnesium alloy. Scr. Mater. **66**, 797–800 (2012). https://doi.org/10.1016/j.scriptamat.2012.02.011
32. Thomas, W., Nicholas, E.: Friction stir welding for the transportation industries. Mater. Des. **18**, 269–273 (1997). https://doi.org/10.1016/s0261-3069(97)00062-9
33. Tutar, M., Aydin, H., Yuce, C., Yavuz, N., Bayram, A.: The optimisation of process parameters for friction stir spot-welded AA3003-H12 aluminium alloy using a Taguchi orthogonal array. Mater. Des. **63**, 789–797 (2014). https://doi.org/10.1016/j.matdes.2014.07.003
34. Yin, Y.H., Sun, N., North, T.H., Hu, S.S.: Influence of tool design on mechanical properties of AZ31 friction stir spot welds. Sci. Technol. Weld. Join. **15**, 81–86 (2010). https://doi.org/10.1179/136217109x12489665059384
35. Yusof, F., Miyashita, Y., Seo, N., Mutoh, Y., Moshwan, R.: Utilising friction spot joining for dissimilar joint between aluminium alloy (A5052) and polyethylene terephthalate. Sci. Technol. Weld. Join. **17**, 544–549 (2012). https://doi.org/10.1179/136217112x13408696326530

Chapter 6
Friction Stir Welding for Joining of Polymers

Debasish Mishra, Santosh K. Sahu, Raju P. Mahto, Surjya K. Pal and Kamal Pal

Abstract The chapter focuses on the welding of thermoplastic polymers. The use of thermoplastics has increased tremendously in the manufacturing industries due to their light-weight characteristic. A detailed study regarding the polymers has been presented and the importance of thermoplastics has also been outlined. The joining technique which has been used in the present work is Friction Stir Welding (FSW). FSW has been one of the major achievements in the field of current welding technologies. Since its invention, the process has been under tremendous research and has been employed to join different metallic alloys of aluminium, magnesium, copper, titanium, etc. The process has also been used to join materials in different joint configurations. Recently, it has been used to weld the thermoplastic materials. An introduction to the FSW technique, the working elements of the process and its constituents have been presented in the chapter. Before the discussion of application of FSW to thermoplastic joining, the other available methods to join thermoplastics such as adhesive bonding and mechanical fastening have been discussed. The literature available with respect to the joining of thermoplastics using FSW has been discussed followed by an experimental study on high density polyethylene (HDPE)

D. Mishra
Advanced Technology Development Centre, Indian Institute of Technology Kharagpur,
Kharagpur 721302, West Bengal, India
e-mail: debsmishra02@gmail.com

S. K. Sahu · K. Pal
Department of Production Engineering, Veer Surendra Sai University of Technology,
Burla 768018, Odisha, India
e-mail: santosh.lenovo@gmail.com

K. Pal
e-mail: kpal5676@gmail.com

R. P. Mahto · S. K. Pal (✉)
Department of Mechanical Engineering, Indian Institute of Technology Kharagpur,
Kharagpur 721302, West Bengal, India
e-mail: skpal@mech.iitkgp.ernet.in

R. P. Mahto
e-mail: rajukec1@gmail.com

© Springer Nature Singapore Pte Ltd. 2019
U. S. Dixit and R. G. Narayanan (eds.), *Strengthening and Joining by Plastic Deformation*, Lecture Notes on Multidisciplinary Industrial Engineering,
https://doi.org/10.1007/978-981-13-0378-4_6

sheets. The results of the study have been presented and the relevant conclusions have been drawn.

Keywords Thermoplastics · Polymer · Joining · Light-weight Friction stir welding

6.1 Introduction

Nowadays, the manufacturers are focusing on developing products which would deliver high performance. One of the attributes for gaining high performance is the light-weighting of the product. With the passage of time, there has been a substantial change in the availability of natural resources worldwide. Various environmental challenges like carbon emission, global warming, etc. have become the topics of concern that needs to be addressed. At present, welding and other fabrication processes are active and highly explored in the manufacturing sectors such as automobile, railways, aerospace, ship building, etc. In order to aid the on-going manufacturing operations, the industries need huge amount of fuel that is being utilised by the equipment in use. Thus, the fuel consumption is increasing day by day owing to the rapid industrialisation, but at the same time, the rate of depletion of these scarce resources is raising the stakes. With the intention to optimise the fuel consumption and to trigger innovation, light-weight and alternative materials and advanced processing techniques that can enhance the efficiency require focused research.

Earlier, steel was the most popular material and was very widely used in different industrial applications. Steel has got high strength and is tough, but is quite heavy. Another material is ceramics, which is strong but brittle. The weight issue emphasises on developing advanced materials which would not only deliver characteristics similar to steel and ceramics but would also enhance the efficiency. The 1910 era led to the development of the very first polymer along with the first light-weight alloys of aluminium [1].

The weight of the products is being reduced by the use of various light-weight materials such as aluminium, magnesium, etc. Thermoplastics, having light-weight have also attracted the manufacturers to employ and explore them in various sectors. The Boeing aircraft, 777 Dreamliner had 50% components manufactured from aluminium while the composite accounted for only 12% [2]. The recent aircraft, 787 Dreamliner contains 50% components manufactured from the composites, which proves that the thermoplastic materials are replacing metallic alloys wherever possible due to their low weight and other unique features [2]. The plastic material can be employed as a single structure and hence can reduce the number of sub-components required in the manufacturing of a particular part. The aerospace industries can be very much driven by the weight factor and these advanced materials help them to reduce the fuel consumption by minimising the net weight of the structure.

Over the years, thermoplastics have proven themselves to be one of the strongest contemporaries, which can not only compete solely but also can combine with other materials to form hybrid structures. The polymeric materials have found their way into daily life and their demand at the present has increased significantly. According to an analysis done by the European Plastic Association, the manufacturing of plastics in the year 2005 was 230 million tonnes and has increased to 322 million tonnes by the end of 2015 [3].

The present chapter aims to present the concept of joining of thermoplastic materials by Friction Stir Welding (FSW). The next section of the chapter is an introduction to the polymeric materials. The importance of the polymers with respect to the metallic alloys and their present applications has been described. The continuing sections explain the traditional techniques of joining polymers namely, the adhesive bonding technique and the mechanical fastening. The merits and the drawbacks of both the techniques have been discussed. Since the main focus of the chapter is to highlight the welding of polymer through FSW, a detailed introduction to the FSW technique has been provided. A case study regarding the experimental analysis of FSW of a thermoplastic has been presented.

6.2   Introduction to Polymers

Polymers are large molecules found in various plastic materials and are formed from small molecular fragments called as monomers. The word poly implies '*many*' and mer implies '*segment*'; mono implies '*one*'. Thus, monomers are the small molecules that are joined together to form a large polymer. The schematic diagram of a polymer is shown in Fig. 6.1. The monomer units are connected through chains.

These monomers are joined together to form a long chain. The process through which the monomers are joined together to form a polymer is termed as polymerisation. The polymers are also termed as 'macromolecules' a scientific term which is translated from Greek meaning '*many units*'. The polymers possess unique properties because of their vast molecular size and average molecular weight spreading from thousands to several million atomic mass units [4]. These macromolecules are composed of atoms bonded together by means of covalent bonds formed by the sharing of electrons. The electrostatic force keeps the individual molecules attracted to one another, but is even weaker than that of the covalent bonds. The properties of anything made from these polymers reflect what is going on at molecular level.

Fig. 6.1 Schematic diagram of a polymer

6.2.1 Structure of Polymer

The polymers are composed of hydrocarbons; compounds of hydrogen and carbon. These carbon atoms are bonded together in large chains. Polyethylene, polypropylene, polystyrene, etc. are the polymers which are composed only of carbon and hydrogen. Polyvinyl chloride has a chlorine atom attached to all the carbon atoms and similarly, Teflon which has a fluorine atom attached to the carbon backbone. The molecular arrangement of a polymer looks like a plate of cooked spaghetti noodles. When the spaghetti has just been cooked and is warm, it takes the shape of the plate and the long strands can slither over one another. As soon as the spaghetti cools down, the strands no longer slither over; instead they stick to each other and become intact. Similarly, the polymer's long chains at moderate temperature can slither over one another and are in a rubbery state. The material is flexible and it does not crack. As the temperature goes down, the long chains stiffen up and they go through a stage known as the glass transition temperature where they become hard and brittle [4].

6.2.2 Characteristics of Polymers

Most of the polymers are lighter in weight. The various characteristics features of polymeric materials which have made them attractive among the manufacturing industries are as follows:

- *Low density*—The polymers are very light in weight and hence are used to manufacture a variety of products ranging from toys to aircraft structures. Some of the polymers can also float in water while others sink.
- *Good corrosion resistance*—Polymers are widely used as packaging bottles which are the storage containers of many acids and other toxic liquids. This shows that how some of them are very highly resistant to chemicals.
- *Good moulding ability*—As compared to the earlier manufacturing ideas where the components were solely made of metals with high strength and weight, the engineering plastics are the current demand and have replaced the metals in many areas. One of the reasons being the moulding ability. The flexibility they offer reduces the components required for a particular part, and hence the maintenance is reduced.
- *Excellent surface finish*—The plastics can be moulded and hence form a neat one component, the surface finish obtained is good. Thus, the dimensional tolerance of the polymers is also high.
- *Economical*—The polymers can be manufactured from different processes. Extrusion process produces pipes, thin fibres, etc. The large body parts of automobiles can be produced from injection moulding process. Some of the polymers may be moulded into different shapes by applying heat directly.

- *Can be produced transparent or in different colours*—The polymer material properties can be enhanced by the addition of different materials and hence their application areas can be broadened.
- *Low surface energy*—The polymers molecules are connected through weak Van der Waal bonds, and hence the intermolecular force of attraction is weak.
- *Low coefficient of friction*—This property is due to the low surface energy of the polymers. This leads to low wettability of the polymers and hence low adhesion property.
- *Poor mechanical properties*—Unlike metals, the polymers lack in tensile strength and also are brittle in nature. Beyond the yield point, the polymers deforms plastically.
- *Poor temperature resistance*—The polymer materials are very sensitive to both high and low temperatures.
- *Thermal and electrical insulators*—The electrical appliances cords and wiring are all covered with polymeric materials, and this idea reinforces the statement that they are electrical insulators. Furthermore, the kitchen utensils like coffee pot, pan handle, etc. are made of polymers which support that the fact that they are thermal insulators too.

6.2.3 Classification of Polymers

The classification of polymers has been depicted in Table 6.1.

6.2.3.1 Natural, Semi-synthetic and Synthetic Polymers

The naturally occurring polymers such as rubber, starch, proteins, etc. are called as the natural polymers. The semi-synthetic polymers such as methyl cellulose, cellulose nitrate, etc. are chemically modified natural polymers. The polymers synthesised in the laboratory are termed as the synthetic polymers. The examples of manmade polymers are polystyrene, polyethylene, polyvinyl alcohol, etc.

Table 6.1 Types of polymer [4]

Basis	Polymer type		
Origin	Natural	Semi-synthetic	Synthetic
Response to heat	Thermoplastics	Thermosetting	
Nature of formation	Addition	Condensation	
Structure	Linear	Branched	Cross linked
Physical properties	Rubber	Plastics	Fibres
Tacticity	Isotactic	Syndiotactic	Atactic

6.2.3.2 Thermoplastics and Thermosetting Polymers

The polymers that can be softened by the application of heat and pressure are termed as thermoplastics. They also can be refabricated. Thermosets on the other hand contain chains that are chemically linked by covalent bonds during the process of polymerization. The thermosets resist deformation, heat softening and hence cannot be processed.

6.2.3.3 Addition and Condensation Polymers

The addition polymers are a result of addition of monomers while the condensation polymers are the results of the reaction between the bi-functional and poly-functional monomer molecules having reactive functional groups. The addition of the monomers occurs in a chain mechanism where they are added with each other in succession to form the addition polymers.

6.2.3.4 Linear, Branched and Cross-linked Polymers

Linear polymers are those in which the monomer units are joined together in a linear fashion. The monomer units when joined together in a branched manner are termed as the branch polymers while if the monomer units are joined in a chain fashion, it is called as cross linked polymers.

6.2.3.5 Rubber, Plastics and Fibres

Rubber is a polymer having a high molecular weight along with long flexible chain and weak intermolecular forces. Plastics are the ones which can be moulded and have high molecular weight. Fibres are long chain polymers with high crystalline regions and possess elasticity lower than that of rubber and plastics.

6.2.3.6 Isotactic, Syndiotactic and Atactic Polymers

The polymers having the characteristic groups in the same side of the main chain are termed as the isotactic polymers. When the groups are arranged in alternative fashion across the main chain, it is called as syndiotactic polymers. The characteristic groups when arranged in irregular fashion across the main chain, the polymer is termed as atactic polymers.

Since, the focus of the present study is the welding of polymers; the chapter concentrates on the thermal response of the polymer. The upcoming section would describe the details of thermoplastic, the difference between thermoplastic and thermosetting polymers, present applications of thermoplastics in manufacturing industries and their future scope.

6.2.4   Introduction to Thermoplastics

The thermoplastic materials when heated above the glass transition temperature become soft while hard when cooled. The thermoplastics can be heated reversibly and solidified in a number of cycles. The structure of thermoplastics is such that it does not undergo any chemical change during the process of heating and forming. However, the recycling process can degrade the colour of the thermoplastic and may bring changes in its properties. The influential properties of thermoplastics which have made it popular are its resistance to chemicals, self-lubrication, high strength, durability, high toughness, etc. Thermosets on the other hand become permanently rigid and hard when heated above the melting temperature. The thermosets have cross-linking chains which makes their structure complex and also prevents the slippage of individual chains. Due to this, when heated, they only result in chemical decomposition. The welding methods cannot be employed to join the thermosets; instead adhesive bonding and mechanical fastening can be used to join them.

The thermoplastics have low weight, high impact resistance, high fracture toughness, excellent anti-corrosion properties, high damage tolerance, design flexibility, low manufacturing and storage cost, etc. Thermoplastics can be described as the three *R's* namely reprocessable, repairable and reformable; which makes them easy to fabricate and ensures cost effectiveness. The thermoplastics are also being combined with metals to form hybrid structures. This metallic part can be explored in sections where strength and stiffness is required while the plastic can be explored where design flexibility is needed. The major advantage of this hybrid structure would be optimal weight thereby improving the overall performance of the equipment [5].

6.2.5   Classification of Thermoplastics

The thermoplastic materials have been broadly classified into two categories which are as follows [1, 4]:

(1) Crystalline
(2) Amorphous

6.2.5.1 Crystalline

This category of thermoplastic polymers is tough, soft and translucent to opaque. They have crystalline regions. The polymers are popular for wear-resistance, bearing and structural applications. The crystalline polymers have better electrical properties, high chemical resistance and low coefficient of friction as compared to the amorphous polymers. The flaws of this category of thermoplastics are the sharp melting point and low impact resistance. The examples include polyethylene, polypropylene, nylon, etc.

6.2.5.2 Amorphous

This polymer is hard, rigid and clear, and has high-dimensional stability. They soften over a range of temperature and hence are easy to thermoform. They have a good impact resistance as compared to that of the crystalline polymers. The molecules are randomly oriented and are tangled with each other. The examples include acrylonitrile butadiene styrene, polycarbonate, polymethyl methacrylate, polyvinyl chloride, etc.

6.2.6 Application of Thermoplastics

Currently, thermoplastics are being widely used in various industries such as automobile, aerospace, construction, etc. The significant applications of thermo-plastics is due to the benefits they offer such as reduction in weight, high specific strength, flexibility in designing, low manufacturing cost, aesthetic properties, etc. The rudder of aircraft A310 has been made out of fibre reinforced plastic. This improvement has led to reduction of weight by 25 and 95% reduction in components by combining parts and forming simple moulded components [6]. The clutch pedal and centre console of Bavarian Motor Works (BMW) has been manufactured from polyamide 66 (PA 66) [7]. The fuselage section and wings of aircraft A380 are manufactured from composites which led to savings in fuel of 17% per passenger [8]. The Boeing Dreamliner aircraft 787 contains 50% components manufactured from composites [2]. Various structures, bolts, window frames for the purpose of construction are manufactured out of thermoplastics.

6.2.7 Additives in Thermoplastics

The engineering thermoplastics are intended to impart high performance and their desirable characteristics such as the mechanical, chemical and physical properties can be modified through the addition of some foreign materials. The additives employed are fillers, plasticizers, flame retardants, colorants, stabilisers, etc. The filler materials may be the fibres, carbon particles, sand, clay, etc. and are intended to improve the tensile and compressive strength of the plastic. With the addition, the plastic becomes more resistant to abrasion and the toughness increases. It also makes the plastic thermally stable so as to withstand high temperatures. The plasticizers are small molecules which occupy the positions between the plastic chains and thus, increase the distance between two units of the plastic. This in turn results in the increase of the flexibility, ductility and the toughness of the plastic. The increment also results in decreasing the hardness and stiffness of the plastic. As discussed earlier, many plastics are damaged when subjected to the ultraviolet rays. The addition of stabilisers helps the plastics to withstand this condition. The colorants impart colour to the plastic for which they are available in a variety of colours giving the product an attractive aesthetic property in which the plastic has been utilised [9].

6.2.8 Classification of Polymer Joining Techniques

The joining of thermoplastics can be broadly divided into three categories, as shown in Fig. 6.2 [10].

6.3 Mechanical Fastening

The polymeric materials are manufactured to the near net shape. The assembly purpose makes use of drilling operation. The mechanical fastening method employs fasteners such as nut, bolts, screws, rivets, etc. to assemble the components. Both similar and dissimilar type materials can be joined by this process [7, 11].

In order to modify the physical, chemical and mechanical properties of the polymers, various types of materials are filled into them. The fibres are one of the crucial components which are very popular and a special class of plastic materials

Fig. 6.2 Classification of polymer joining methods

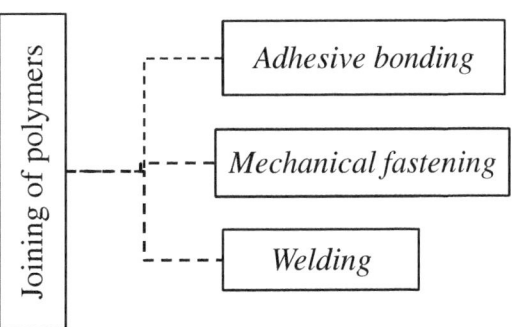

called the fibre reinforced plastics or commonly known as the FRPs are manufactured when the plastics are filled with fibre. The drilling operations when carried
out on the plastic material cuts off these fibres aligned in a particular fashion and
thus results in decreasing the strength of the component. The holes also lead to
stress concentration. The fastener head has to be chosen carefully since they produce undesirable effects in the assembled component. Flat washers are always
recommended to be in use because they help to distribute the force uniformly in the
mating components. Thread lockers are used sometimes to secure the fasteners. The
locker has to be again selected with care because they can be chemically aggressive
to the plastic components.

The total component design is very important. Plastics expand more and are
highly sensitive to change in temperature. The plastic when exposed to high temperature, their stiffness decreases. With further increase in temperature, the part
buckles. The opposite effect is equally dangerous; with decrease in the temperature,
the plastic becomes stiffer. This whole phenomenon is crucial for the mating
components as it would ultimately result in failure of the component. While
designing the dissimilar polymer joining, the properties of both the plastics have to
be taken into account.

6.3.1 Advantages of Mechanical Fastening

The following are the merits of mechanical fastening method:

- The operation requires only drilling operation followed by fastening the sub-
  strates with the fastener and thus, the total operation is very simple and less time
  consuming.
- No surface preparation is required except cleaning the hole and the substrates.
- Dissimilar plastics can also be joined.
- The joint can be easily reopened and maintenance activities as required can be
  carried out.
- The integrity of the joint can be predicted.

6.3.2 Drawbacks of Mechanical Fastening

The following are the drawbacks of the mechanical fastening method [11]:

- The inclusion of fasteners into the parts to achieve the joint increases the weight.
- The holes drilled in the substrate leads to concentration of stress.
- The fasteners also develop stress with the passage of time which makes the bond
  weak.

- The mating part when used in rotating equipment may lead to loosening of the fastener and thus may result in failure. The loosening may also happen due to relaxation in the stress and environmental factor such as moisture.
- The mechanical fastening method may not completely produce a leak proof joint.
- Plastic surface may disrupt due to improper fastening of the screw or rivet.

6.4   Adhesive Bonding

Adhesives are materials that are used to hold two surfaces together. The adhesive must be able to wet the surface on which it is applied and adhere to it [12, 13]. The preparation of an adhesively bonded joint can be carried out through two main operations; one is the design and the second is the manufacturing. The events which come under design and manufacturing have been shown in the Fig. 6.3 [14].

The major characteristics of an adhesive are as follows:

- *Make the surface wet*—It must flow all over the surface on which they are being applied in order to remove all the air and other contaminants present there.
- *Adhere to it*—The adhesive should start to adhere after flowing over the whole area, must stay in the same area and become tacky.
- *Develop strength*—Adhesive must change its structure to become non-tacky and strong.
- *Remain stable*—The bond must not be prone to environmental changes and age.

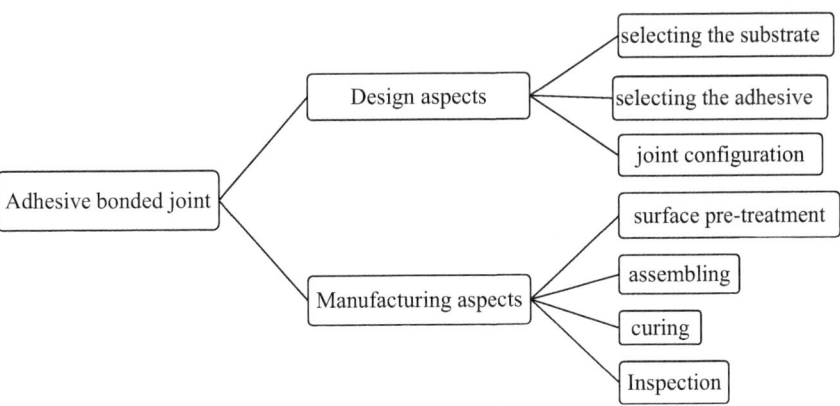

Fig. 6.3 Events in the making of an adhesive bonded joint

6.4.1 Adhesion and Cohesion

The adhesion is an interfacial phenomenon. It is the bonding between two materials namely an adhesive and a substrate. The cohesion is the internal strength of the adhesive which is a result of the various interactions between the molecules of the adhesive. There are three main theories of adhesive namely the adsorption theory, electrical theory and the diffusion theory. The adsorption is an effect of the surface energy. The process of adsorption is either physiosorption which is a characteristic of the Van der Waal forces or it is chemisorption, a characteristic of covalent bonds. The atoms in the adsorbent are not totally surrounded by the other atoms in the adsorbent and hence there would be attraction force from the atoms in the adsorbate. The electrical theory explains the electrostatic force at the interface of the bond. The adhesion bonding is a result of the columbic attraction between the two surfaces. The diffusion theory explains adhesion as a result of the interdiffusion of molecules between the surface and the adhesive.

6.4.2 Bonding Zones

The three bonding zones are namely; adhesive, transition and the cohesion zone. The adhesion is caused due to the molecular interactions between the substrate and the adhesive. This interaction can be a strong chemical bond or just a weak intermolecular attraction. But the chemical bonds are not formed everywhere. Some typical combinations like polyurethane and glass, silicone and glass, etc. have the chemical bonds in between them. These joints are strong and its durability depends upon the resistance of the bond to moisture. In apropos to this chemical bond, the micro-mechanical bonding also plays an important role. The irregularities in the surface of the substrate act as areas which allow the adhesive to flow in and in turn it increases the bond strength. The adhesive clings mechanically to the roughened surface. The transition zone refers to the area where the various properties of the adhesive such as mechanical, chemical, etc. are altered. It is particularly important for the thin bonds, where the joint characteristics are evaluated by the properties of the transition zone. The cohesion zone involves the solidification of the adhesive through the bonds formed between the adhesive molecules.

6.4.3 Bonding Fundamentals

As discussed in the previous sections, the bonding takes place in three different modes namely; the Van der Waal bonding which are weak, the polar group interactions which are comparatively stronger than that of the Van der Waal bonds and lastly the chemical bonds which are the strongest bonds. We already have discussed that the adhesive must wet the surface in order to adhere properly. If the polarity of any surface is low, then the adhesive cannot wet the surface easily. Most of the plastics have low polarity and hence it is necessary to apply primer in order to extract the most out of the adhesive [15].

6.4.4 Surface Preparation

The thermoplastics with low energy mostly contain carbon–hydrogen or carbon–fluorine bonds because of which they have low polarity. The higher energy thermoplastics have other constituents like nitrogen or oxygen in abundance which increases their polarity. Thus, the plastics require surface preparation. The various methods that are employed are corona treatment, flame treatment, plasma treatment, etc. These methods make use of gases which are impinged on the surface under oxidising conditions to create surface polarity [16].

6.4.5 Types of Adhesives

The adhesives in the market are available in wide range and they can be classified on the basis of physical change they undergo, chemical reactions involved, typical application areas, etc. The major categories of adhesive have been explained below

- *Anaerobics*—This adhesive has the anaerobic curing property. Once they are away from the air, they will form solid polymer immediately. It consists of a monomer and a catalyst for bonding. The name is anaerobic since the adhesive does not require oxygen during its process of curing. The typical advantages of this type of adhesive is that they cure at room temperature, offer good strength, have mild odour, etc. But the adhesive cannot be used for joining plastics as it is driven by air during the curing process that can result in increasing the process time, requirements for storing and handling the adhesive, etc. [17].
- *Cynoacrylates*—This adhesive is known for its extreme low curing time. It imparts good bond strength and also it eliminates the requirement of any external pressure but such low drying time makes the bond impossible to break once formed. The adhesive needs to be handled very carefully since it may damage the fingers and also may cause irritation. Again, the storage is also a concern since once the lid of the container is opened; the total content may dry up. The adhesive has low resistance to heat and also has very low impact resistance [16].
- *Epoxies*—This class of adhesive consists of two parts; a resin and a hardener, which are mixed together before use. They impart good bond strength but they

are toxic and flammable. They need long curing time, thus, increasing the process time. Extreme care is needed while handling and storing them.

- **Silicones**—The silicones are available both in one part and two-part systems. They impart good strength and flexibility. Dissimilar materials can be joined by using them. They liberate gases while curing. They offer good resistance to high temperature and chemicals.
- **Acrylics**—This class of adhesive can be either water-based or solvent-based. The water-based acrylics have lower curing time as compared to the solvent based but the solvent-based are resistant to other solvents, chemicals and water. The acrylics are sub-divided into two namely; pure and modified acrylics. The pure acrylics have a low stickiness and less adhesion as compared to the modified acrylics.
- **Rubber**—The rubber adhesives are based on the latexes solution and solidify though the loss of the solvent medium. They are not suitable for sustained loadings. They can be sub-divided into two; natural and synthetic rubber. The natural rubber is used as surface protection tapes and they possess higher stickiness than that of the acrylics.

6.4.6 Demerits of Adhesive Bonding

Although, the adhesive bonding technology is flexible, offers good bond strength and in some cases less expensive, it has numerous disadvantages which makes it less suitable for industrial applications [6]. The drawbacks of using adhesives have been described below.

- **Curing time**—The strength to the structure is not achieved immediately. Some adhesives offer fast curing but most of them have slow curing time. Again, this curing time is also affected by the environmental conditions. Hence, the process time is raised.
- **Resistance to temperature**—Most of the adhesives are polymer-based and thus have low resistance to temperature. Thus, the structures obtained through adhesive cannot be deployed in areas where high temperatures are involved.
- **Ageing**—The adhesively bonded joint's strength is highly affected by the physical and chemical actions going on in the environment such as the presence of moisture, ultraviolet rays, chemical attacks, etc. There are adhesives which are resistant from the ultraviolet rays attack but many of them break down under the ray's action. The solution to the problem would be the selection of the adhesive according to the environmental conditions in which the bonded joint has to be deployed. This in turn would lead to undertake a number of quality tests to ensure the effectiveness of the joint.
- **Surface preparation**—For achieving good bond strength, the surface preparation has to be carried out. Preparing the surface prior to application of the adhesive ensures proper adhesion between the substrates and the adhesive.

- *Life of the joint*—The non-destructive techniques cannot be employed in the adhesively bonded joints to evaluate the strength. Thus, predicting the life of the joint is not possible and hence can be dangerous when the joint is used in sophisticated equipment.
- *Safety and environment*—Almost all the adhesives are needed to be handled very carefully since if they stick to fingers they may damage them. Also, during the process of curing some of the adhesives liberate toxic gases which are again harmful for the person who are involved in handling them. The adhesive container once opened has to be stored and handled with proper care. The waste produced during the process of application requires attention for cautious handling for recycling and treatment.

6.5 Introduction to FSW

The FSW is a solid state friction welding method which makes use of a specialised tool creating frictional heat during the process. It was invented at The Welding Institute in the year 1991, Cambridge, United Kingdom [18]. The tool used is not consumed in the process rather its geometry and dimension are altered in combinations which could contribute to the process of heat generation. The process can be employed to any joint configuration; but has been widely applied to butt and lap configurations since its invention. The process was developed for the welding of aluminium alloys, but has been extensively applied to other materials like magnesium, steel, titanium, copper, thermoplastics, composites etc. as well [19, 20]. The schematic view of the process is shown in Fig. 6.4.

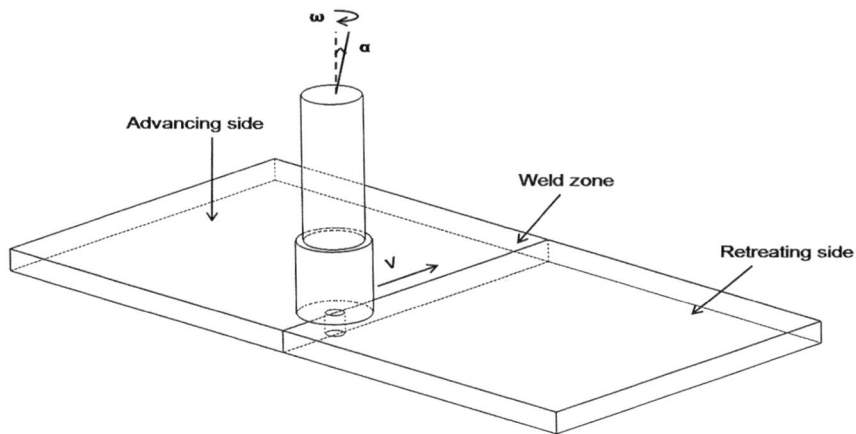

Fig. 6.4 Schematic view of FSW

6.5.1 FSW Tool

The specialised tool used in FSW consists of two features called as the pin and the shoulder. A typical FSW tool has been shown in Fig. 6.5. The tool for the welding are fabricated from a variety of materials which solely depends upon the selected work-pieces. The diameter of shoulder to the diameter of pin is kept in the ratio of 3:1 or 4:1 [21]. The length of the pin is kept 0.3 or 0.4 mm less than that of the thickness of the work-piece. The various tool materials that have been used so far in FSW have been tabulated in Table 6.2. The first FSW tool used had its pin and shoulder both of cylindrical shape. With the passage of time, pin and shoulder with varying features such as conical, square, tri-flute, etc. have been developed to investigate its effect on the welding process.

6.5.2 Principle of FSW

The welding technique occurs sequentially in three stages namely; plunging, traversing and retracting. The tool is fixed in the machine spindle. The work-pieces to be weld are tightly fixed on to the machine bed so as to prevent the edges in contact to tear apart. The work-pieces are maintained at zero root gap. The tool rotates at an angular velocity which is termed as the rotational speed. The tool travels over the substrates and is referred as the traverse speed. The side of the weld where the rotational vector direction and travel vector direction are same is termed as the advancing side (AS) and the other side for which they are opposite is termed as the retreating side (RS). The various forces involved in FSW are the translational force acting in X-direction, transverse force in Y-direction and the axial force in the

Fig. 6.5 FSW tool

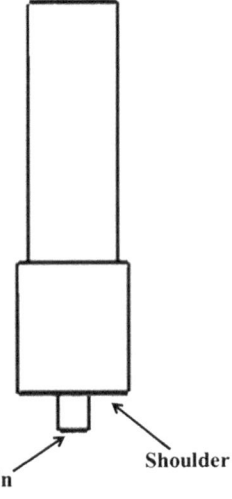

Pin   Shoulder

Table 6.2 Various tool materials

S. No.	Tool material
1.	H13
2.	Tungsten carbide
3.	PCBN
4.	SAE 1050 steel
5.	Stellite 6B
6.	MP-159
7.	Ferro-Tic-HT-6A
8.	Mo-TZM
9.	Rhenium
10.	Ferro-TiC Sk

H13 chromium-molybdenum steel, *PCBN* poly-cubic boron nitride, *Stellite 6B* cobalt based alloy, *MP-159* nickel cobalt based alloy, *Ferro-Tic-HT-6A* steel alloy bonded titanium carbide; non-magnetic and wear resistance, *Mo-TZM* titanium zirconium molybdenum, *Ferro-TiC* steel alloy bonded titanium carbide

Z-direction. The moment about the axis of rotation is referred as the torque. The tool can also be provided a tilt for better plunging and forging of the softened material and is termed as the tilt angle. Plunge depth is also provided to the tool to ensure that the tool has properly plunged into the work-piece. It is specified as the depth to which the shoulder sinks into the work-piece [22, 23].

The rotating tool slowly plunges into the materials as the pin initially makes a contact with the work-piece. The contact of pin with work-piece results in the generation of frictional heat. After the pin, the shoulder makes a contact with the work-piece and again contributes to the process of heat generation. These events are parts of plunging stage and are referred in Fig. 6.6a–c. The frictional heat generated deforms the material plastically and softens it. The tool brings the material to a stage where it can flow. Once the tool has completely plunged, it starts traversing over the joint line dragging the soft material from the AS through the RS and then again forging it back in the AS. At the end of joint line, the tool retracts out from the joint leaving a key-hole which is referred as the retraction, as shown in Fig. 6.6d.

The various parameters involved in FSW can be categorised into three categories namely; welding, material and design parameters. Tool rotational speed (ω), traverse speed (v), tilt angle (α) and plunge depth (d) are the attributes of welding parameters. The combinations of ω and v are of prime importance in the process of welding. A higher ω and lower v; would result in hotter welds leading to defects. Similarly, a lower ω and higher v would lack in heat required and again result in defective welds. A higher α would make the tool plunge more and hence would also result in defective welds.

The total heat generated during the process can be explained using the following equation [24]:

- Rigid clamping system is required to prevent the substrate tearing apart.
- A key-hole is formed in the weld zone at the end as the tool retracts out from the joint line.
- Backing plate is required during the welding.
- The initial investment cost is very high.

6.5.4   Application of FSW Process

FSW process has been widely adopted in various industrial sectors like aerospace, marine, railways, automobiles, construction, ship building, etc. Some of the industries which are using FSW to in their production process are Apple, Honda, Boeing, Ford, Hitachi, etc. The iMac utilised FSW to weld the front and the back panel having a thickness of 5 mm [30]. Honda Accord has its front panel manufactured from FSW of steel and aluminium [31]. The dissimilar material welding is very popular since it reduces the weight. The aluminium and steel dissimilar combination is being used in aerospace and automobile industries to build parts such as bumpers, pillars, chassis, etc. [32]. Similarly, the aluminium and magnesium combination welds are used for parts such as clutch, transmission exhaust décor, etc. [32]. The FSW has been applied successfully in welding of aluminium to copper which has found applications in manufacturing of household utensils and industrial power protection [33–36]. Other dissimilar combinations such as magnesium with copper and aluminium with titanium are also being explored in the aerospace and automobile sectors [37, 38]. The Fosen Mek's Cruise ship 'The World' contains decks manufactured by FSW [39]. The commuter EMU series 20,000 manufactured by Hitachi contains FSW roof panels. The 700 series Shinkansen rail contains aluminium floor panels manufactured by FSW [40]. The Ford GT contains the centre panel manufactured from aluminium fabricated by FSW [41].

6.5.5   Defects in FSW

The combinations of the process parameters are crucial as it can only lead to generation of defect-free welds. The various defects that may arise during the process are given in Fig. 6.7 [42].

The tunnel defect is mostly found in the welds with too high traverse rates. The higher the welding rate, the lower would be the generation of heat during the welding process. Thus, the materials to be welded would not be able to reach the plastic stage and hence it cannot be stirred which would result in improper mixing. It is found near the AS. The kissing bond defect occurs in the welds as a result of improper fusion between the two base materials. There is no metallurgical bond

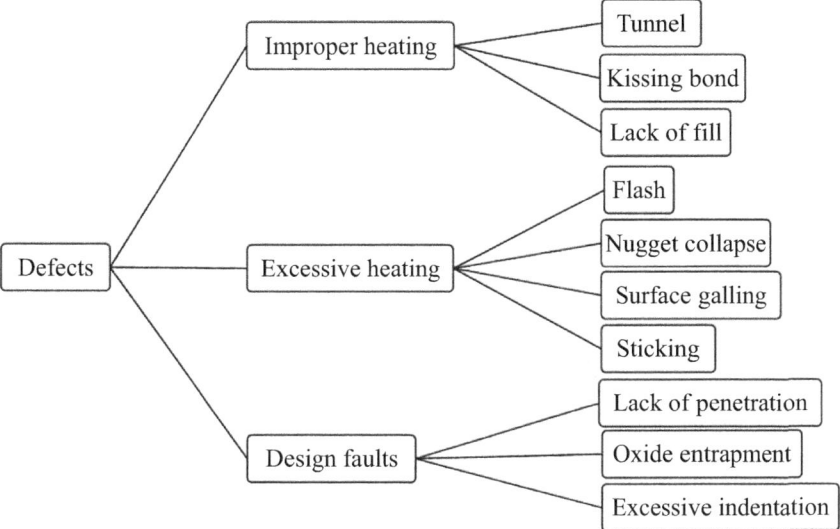

Fig. 6.7 Defects in FSW

achieved in this case. The lack of fill is categorised as a surface defect which occurs due to insufficient plunge force during the process of welding.

In welds produced with high rotational speeds and low welding speeds, the heat generation is at the higher side which leads to excessive softening of the base materials. The nugget collapses as a result of this. This defect can be eliminated by controlling the heat development in the process. Also, as a result of this high heat, the plasticized material is sometimes flow out from the shoulder zone, and hence the material does not go into the weld zone, rather it gets accumulated in the RS. This accumulation of the material is termed as the flash. The root defects in FSW are a result of higher length of the pin which softens the material to higher extent and thus, the material at the bottom sticks to the bed. Surface galling refers to the voids seen in the AS due to excessive heat.

The oxide entrapment is a problem with metals such as aluminium which have a high affinity for oxygen. The remedial to this is the preparation of the material prior to welding; removal of the oxide layer. The lack of penetration is the result of smaller pin length which fails to achieve the required homogenization. Extreme indentation is caused due to the higher plunge depth. This further leads to loss of the material from the weld zone.

6.5.6 FSW Metallurgy

A typical joint fabricated by FSW contains three weld zones namely; nugget zone (NZ), thermo-mechanically affected zone (TMAZ), heat affected zone (HAZ) and the parent/base metal zone (BM). The centre of the weld is referred as the NZ and it consists of refined equiaxed grain structure due to the dynamic recrystallization. The width of the zone depends upon the parameters selected. The area near by NZ is TMAZ. This zone does not experience as much change as NZ but is affected both mechanically and thermally. It consists of partially recrystallized grains. The adjacent area to TMAZ is referred as the HAZ which is only affected by the heat during the process of welding. The BM is the base metal which has properties same as that before the process of welding [23, 43, 44].

6.6 Literature Related to FSW of Thermoplastics

The first work in this regard was a comparative analysis between the various available techniques to join thermoplastics. The material under the test was polypropylene (PP) [45]. The various methods employed to join the thermoplastic sheets were ultrasonic, hot gas, hot plate, extrusion, friction, adhesive and FSW. The aim of the research work was to investigate and analyse the scope of FSW to weld thermoplastics with respect to other methods. A comparative study was performed in terms of energy required by a process and the total cost involved in it. Ultrasonic method requires energy directors such as cone shaped protrusions which would flow and fuse into the joint, tight fit to achieve frictional energy between mating parts, etc. clean surface is needed for adhesive bonding, hot gas and extrusion process require groove, friction method requires flatten surfaces whereas the hot plate and FSW needs no surface preparation. The cost requirement by adhesive is the least among all but, as discussed earlier, the method has got various demerits which lack quality. The joint efficiency achieved with FSW was 95% and also the repeatability of FSW is high. The potential of FSW was also studied on 15 mm thick PP sheets [46]. The effect of rotational and welding speed was also studied. The joint strength efficiency obtained was 50% of that of the base material strength. The authors concluded that the use of grooves in the tool is necessary which would collect the softened polymeric material with it and forge it back in the joint line. TWI reported the successful welding of 9 mm thick PP sheets with a new variant of FSW called as the *Viblade* welding [47]. The frictional heat was generated by the blade vibrating in a reciprocating manner parallel to the joint line. The traditional FSW tools tends to damage the surface of the thermoplastic material since, they are softer than the metals. For this purpose, various new tools were developed and patented to be used for welding thermoplastics by FSW [48]. Still, numerous works have been carried out by using the traditional FSW tools to weld the thermoplastic materials. The upcoming sub-sections present the available literature in the context.

6.6.1 FSW Tools for Polymers

The tool components, i.e. pin and the shoulder both are responsible for the generation of heat by friction. Thermoplastics are softer in nature with low melting point. They also have low binding energy. The pin rotation is very much crucial in order to make the softened material flow from AS to RS, whereas the shoulder rotation has proved to be damaging. It ruptures the fibres present on the surface; thus, degrades the quality of the parent material. A heated shoe tool was employed to weld the PP plates [49]. The long shoe with a pin was responsible for heating and exerting force over a larger area of the PP plates. A coil was placed inside the shoe which provided the heat during the welding operation. The prime goal of the study was to evaluate the effect of FSW on polymer microstructure. The shoulder part in traditional FSW tool has been replaced by this long shoe. Thus, the shoulder is stationary and it prevents the softened material to be expelled from the weld zone. The author concluded that larger pin diameter, low feed rate and high shoe temperature were the best conditions for obtaining good quality welds. The stationary shoulder tool was again used to weld acrylonitrile butadiene styrene (ABS) plates [50]. A conical threaded pin was used. No external heating in this case was provided, but still the welds achieved smooth surfaces. Due to the negligible thermal conductivity of the polymers, the heat was not conducted from the AS to the RS. Thus, the temperatures in AS and RS were non-uniform. A new variant of tool to weld the polymers was further developed which consisted of two stationary shoulders made of polytetrafluoroethylene (PTFE) [51]. The pin was tightened in between the two shoulders. The objective was to eliminate the root flaws caused by the traditional FSW tools. The material under the experimental analysis was ABS. Two different pin profiles were used during the analysis namely; cylindrical pin and convex pin. The authors observed that the welds produced with convex pin had higher tensile strength than that of the welds produced by using cylindrical pin. It was because of the pin design; convex pin had more area than the cylindrical pin and hence it was able to create sufficient amount of heat required to plasticize the work-piece. Moreover, the upper and bottom shoulder surfaces prevented the expulsion of the softened material which strengthened the joint. The shoulder also eliminated the root defects and back slit. In another research work, two tools with left handed threaded pin were employed to weld the Nylon-6 plates [52]. The tools were fed in clockwise and anti-clockwise directions separately. The objective was to evaluate the effect of thread with direction in which it is fed on the material flow path. The threads were observed providing a path for the softened polymer material to flow along the edges of the tool. When the tool was fed in clockwise direction; the flutes ran from bottom to the top, and as the pin rotation was also in the clockwise direction, it dragged and threw the softened material out from the weld zone. But when fed in anti-clockwise direction, the flute and pin rotation direction were opposite to each other and this made the material flow evenly on the weld region, thus, good quality weld was achieved. The PP plates with 20% carbon fibres were under investigation for welding by FSW with four different pin profiles namely; threaded cylindrical, threaded cylindrical-conical, simple cylindrical-conical and threaded conical [53]. It was observed that the conical pin was not suitable in design as it is unable to achieve the homogenization of the softened material. The pin instead created a void in the stir zone.

6.6.2 Process Parameters

As discussed earlier, the various process parameters in FSW are the tool rotational speed (ω), tool traverse speed (v), tool tilt angle (α) and the plunge depth (d). Most of the research works have been carried out to identify the effect of ω, v and α. Thermoplastics materials differ significantly from the metals with respect to the intermolecular energy, melting point, thermal conductivity, coefficient of thermal expansion, etc. As such, the process parameters for both also would vary. A significant higher ω would generate high heat which would completely destroy the polymeric material by melting it whereas a lower ω would result in insufficient heat and hence cannot achieve plastic deformation [54]. The effect of tool tilt angle was studied and it was found that it provided assistance in achieving the required heat for the process [55–57]. But at the same time, a higher tilting of the tool damaged the work-piece. The ω was found to be the most influencing parameter having a contribution of 73.85% among others [58]. The heating effect while employing stationary shoulder was also found to be important. With heating less than 90 °C, material flow was non-uniform while when it was above 130 °C, defects like flash and tunnel were observed [55]. The polyethylene sheets were welded with double passes in order to eliminate the root defects [56]. The ultrahigh molecular weight polyethylene (UHMW-PE) sheets were preheated before welding to reduce the effect of non-uniform heat distribution during the welding process [57].

6.6.3 Achieved Joint Strength

The joint strength in a welding operation is a prime feature as it ensures the efficiency of the process to joint two materials. In case of FSW of thermoplastics, some works have obtained joint strength efficiency of 95% [49] while others have only 25% [59]. Table 6.3 presents the achieved joint strength efficiency, the material under experiment, process parameters selected and the tool employed in various works.

Table 6.3 Joint strength achieved in FSW of thermoplastics

Base material	Tool employed	Parameters (ω, v, α T, F)	Joint strength efficiency (%)	Ref.
HDPE	Cylindrical pin and shoulder	$\omega = 3000$–$20,000$; $v = 10$–44	88	[62]
UHMW-PE	Cylindrical shoulder with threaded pin	$\omega = 960$ and 1960; $v = 10$ and 20	89	[57]
PP with 30% GF	Taper pin with groove, triangle pin with thread, trianglular pin and cylindrical pin with groove	$\omega = 400$–1000; $v = 8$–20; $\alpha = 0{,}1$	25	[59]
PP with 20% CF	Threaded cylindrical, threaded cylindrical-conical, conical, threaded conical	$\omega = 1000$; $v = 16$; $\alpha = 1$	19	[53]
PP	*Hot shoe* (stationary shoulder)	$\omega = 1080$; $v = 51$, 102, 203, 305; $T = 110$, 127, 143, 160, 177	95	[49]
Nylon-6	Two left handed threaded tools; (fed in clockwise and anti-clockwise direction)	$\omega = 1000$, $v = 10$	19.7 (clockwise); 47.4 (anti-clockwise)	[52]
ABS	*Hot shoe* (stationary shoulder)	$\omega = 800$–1600; $v = 20$–80, $T = 50$–100	88	[63]
PC	Cylindrical shoulder with threaded pin	$\omega = 1000$–1850; $v = 20$ and 40; $\alpha = 1$ and 3	30.78	[64]
ABS	Stationery shoulder with conical pin	$\omega = 1000$–1500; $v = 50$–200, $F = 0.75$–4	67	[50]
ABS	Convex pin, simple pin with two stationary shoulders	$\omega = 400$ –800; $v = 20$ to 60	60.63	[51]
MDPE	Cylindrical pin and shoulder	$\omega = 1000$–1800; $v = 12$, 16, 20; $\alpha = 1$ and 2	75	[65]

T tool temperature in °C; F axial force in kN

Joint strength efficiency = (obtained tensile strength/base material tensile strength) × 100%

HDPE high density polyethylene, *UHMW-PE* ultrahigh molecular weight polyethylene, *PP* polypropylene, *GF* glass fibre, *CF* carbon fibre, *ABS* acrylonitrile butadiene styrene, *PC* polycarbonate, *MDPE* medium density polyethylene

6.6.4 Defects in Welding Thermoplastics

In prior discussion it was mentioned that FSW has unique defects as compared to the traditional welding methods. The root defect was observed while welding the polymeric materials with a conical pin [49]. This defect was due to the conical design which was unable to create the required turbulence. If a higher pin length is used, the work-piece material at the bottom will stick to the bottom, while a pin with shorter length as compared to the work-piece thickness would result in lack of penetration. The peeling defect was observed in many research works while welding using traditional FSW tools [60]. Flash was observed while welding PP with 20% carbon fibre in a lap configuration [53]. The flash generally occurs due to higher plunging of the tool into the work-piece. The wormhole defect was observed in the AS of the welded samples which occurred due to the loss of material from the weld zone [59].

6.7 A Case Study

This section presents an experimental analysis carried out to evaluate FSW of HDPE sheets. The HDPE polymer is a versatile thermoplastic. It is being employed in many applications. The material has the ability to be recycled in both its rigid and flexible forms. It possesses the present day requirement of manufacturing industries —light-weight and superior strength. The thermoplastic is employed by various automobile manufacturing industries hence forth to reduce their equipment's weight. It has got good impact resistance and is also resistant to any chemical attack. Thus, HDPE has been chosen as the research material for the present analysis. Moreover, it also has been the most selected thermoplastic material for research in the field of FSW. The objective of the experimental analysis was to evaluate the effect of process parameter on the welded samples.

6.7.1 Materials, Methods and Machineries

The HDPE specimens with dimensions of 100 mm × 100 mm × 6 mm were prepared for the purpose of welding. The welding was performed in butt configuration. The physical properties of the base material have been tabulated in Table 6.4.

The tool used for welding was fabricated from H13 steel and its dimensions have been shown in Fig. 6.8. The tool shoulder and pin were made cylindrical. The tool had a shoulder diameter (D) of 14, 5 and 5.6 mm being the diameter (d) and height (h) of the pin, respectively.

Table 6.4 Physical properties of HDPE

Thermal conductivity (W/m °C)	0.46
Melting point (°C)	126
Density (kg/m^3)	950
Coefficient of thermal expansion $\times 10^{-6}$ °C^{-1}	100–200
Ultimate tensile strength (MPa)	33

Fig. 6.8 FSW tool used for experiments

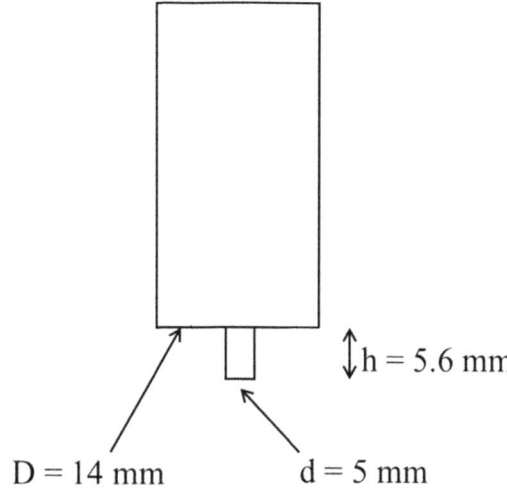

$h = 5.6$ mm

$D = 14$ mm $d = 5$ mm

A 2 tonne NC controlled FSW machine manufactured by ETA Technology, Bangalore was utilised for the FSW. The set-up has been shown in Fig. 6.9. It is a linear welding machine with a maximum spindle rotation of 3000 rpm, maximum welding speed of 1000 mm/min and a tilt angle of ±10°.

The welding parameters selected for the experiment have been shown in Table 6.5. The tool rotational speed and the traverse speed were varied while a constant tilt angle of 1° and plunge depth of 0.1 mm were applied to all the experimental runs to ensure better plunging of the tool in the work-piece. The samples to be welded were tightened properly on the bed with the help of the fixture, and a zero gap was maintained. The tool fixed in the spindle was then traversed over the joint line to ensure that there was no mismatch between the joint line and the face of the pin. Since a tilt of 1° was applied, the tool plunge value was recorded by making contact between the trailing edge of the tool and the work-piece surface. This would ensure better mixing of the plasticized material and would help to minimise the formation of defects. If the plunge would have been measured with the leading edge of the tool, then the trailing edge would have plunged more resulting in excessive material loss from the weld zone.

The welded samples were removed after 2 min from the machine bed to avoid distortion of the sample from the fixed ends. The tensile specimens were then cut from the welded samples as per the standard ASTM D638. A typical tensile test

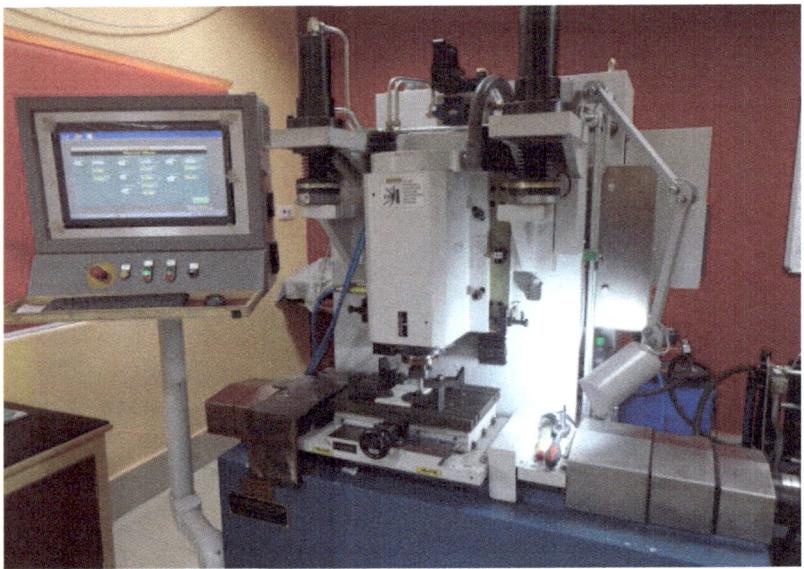

Fig. 6.9 FSW machine set-up

Table 6.5 Experimental parameters

Experiment run	Tool rotational speed (ω, rpm)	Tool traverse speed (v, mm/min)	Tool tilt angle (α, °)
1	500	10	1
2	600	20	1
3	800	30	1

sample which is in accordance with the above standard is shown in Fig. 6.10. The tensile testing was carried out in a universal tensile testing machine (Instron, 1344).

6.7.2 Observations

During the trial experiments, it was observed that with ω less than 500 rpm, the heat generated was insufficient to plastically deform the material and as such the joint strength was poor. When the ω was kept above 900 rpm, the heat developed was very high. In both the cases, the welding speed selected was 10 mm/min. Similarly, when the welding speed was kept less than 10 mm/min, the generated heat almost melts the material under the shoulder surface. Also, low welding speed would reduce the process efficiency as they would need more time to complete the

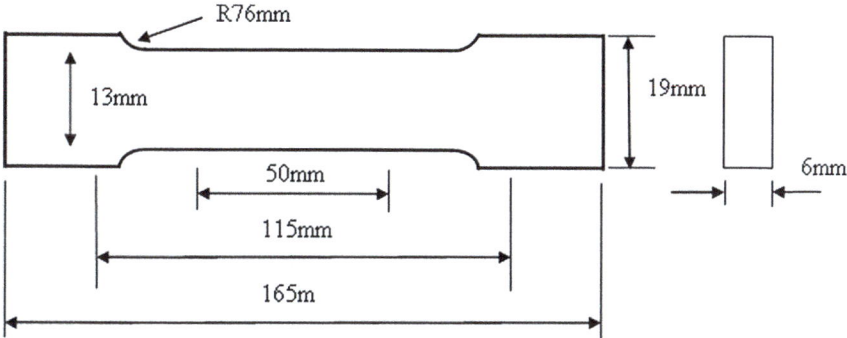

Fig. 6.10 Tensile test specimen dimension

process. It was also seen that with an increase of tool rotational speed, the material peeling rate also increased. This was because of the higher frictional rate which disrupted the thermoplastic surface and initiated the fibre breakage. The material peeling with different tool rotational speeds has been compared visually in Fig. 6.11. When welding was performed with 500 and 900 rpm, the peeling was more in the latter case. The tool dimensions and other parameters were kept same for both.

Also during the trial experiments, two tools with various shoulder diameters, 24 and 18 mm, were selected to determine their effect on the welded samples. The pin diameter selected was 6 mm. It was found visually that the stirred volume of the plasticized material increased with the increase in diameter of the shoulder. With 24 mm diameter, the heat generation rate was also more. Loss of the material from the weld zone was observed due to this high amount of deformation and friction. The sample welded with 24 mm shoulder diameter has been shown in Fig. 6.12. This result was also observed with 18 mm shoulder diameter which shows that higher shoulder diameter is devastating for thermoplastics. One more observation

(a)   **(b)**

Fig. 6.11 Material peeling. **a** 500 rpm, **b** 900 rpm

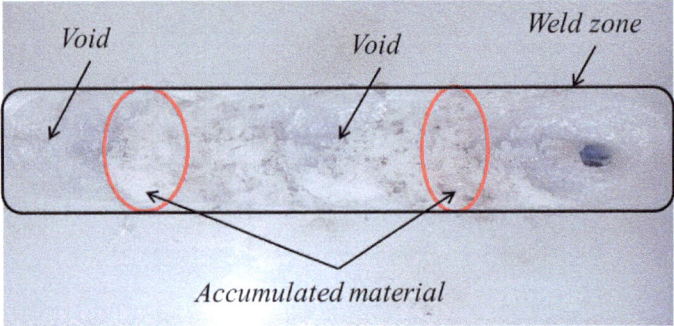

Fig. 6.12 Effect of 24 mm shoulder diameter

should also be noted that the shoulder rotation is also harmful for thermoplastics since it is resulting in breakage of the fibre on the material's surface and hence destroys the material beneath the influence of the tool.

Thus, to prevent the loss of material from the weld zone, it was necessary to reduce the dimensions of the tool. Hence, the tool shoulder diameter was set as 14 mm and that of the pin as 5 mm.

6.7.2.1 Physical Appearance of the Welded Sample

The cylindrical tool and the selected process parameters successfully welded the HDPE sheets. An appearance of the sample welded at $\omega = 600$ rpm and $v = 10$ mm/min is shown in Fig. 6.13. The first image is the front side of the welded sample. The weld zone seems to be regular and the surface finish was good. Peeling of the materials during the experiment with these parameters was also observed, but the volume of material removal was low. The second image is the back view of the sample where the flow of material can be seen in the form of concentric rings. Also, the width of the weld zone is more in the front side as compared to the back which shows that the upper surface is more influenced by the shoulder, while the bottom is the pin influenced area. The third figure is an enlarged view of the front side of the welded part. Although, the surface looked smooth, various tiny voids were there. These voids were present as a result of the peeling defect. Thus, for thermoplastic materials, it is better to avoid the rotation of the shoulder.

6.7.2.2 Variation of Axial Load with Time

Since, the thermoplastics have properties very different from that of the metals, it was necessary to study the variation of axial welding force during the process of welding. The variation of the force for a particular welded sample ($\omega = 800$ rpm,

(a)

(b)

(c)

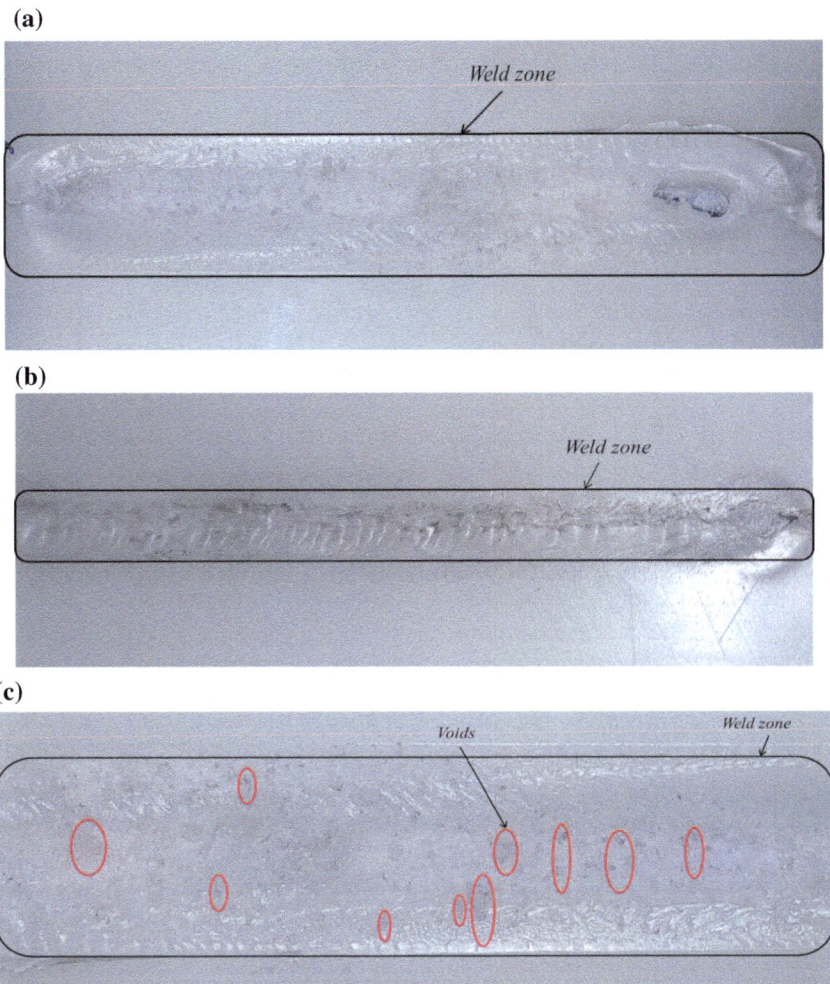

Fig. 6.13 Welded sample appearance. **a** Front view, **b** back view, **c** enlarged view of the weld zone

$v = 20$ mm/min) is shown in Fig. 6.14. The first peak observed in the figure shows the contact of the pin with the base materials. The force started to rise as the material was at ambient temperature condition. With further plunging of the pin, the material was deformed plastically and then a drop in the force can be observed. The second peak is a result of the initial contact of the shoulder with the work-piece. The force started to rise and attained a peak value of 2.8 kN. The peak axial load in case of FSW of aluminium is approximately 8.9 kN [61], thus, this shows that the force required in case of thermoplastics is low. At this point of time, a sudden drop in the signal is observed and is termed as the dwelling stage. This accounted for

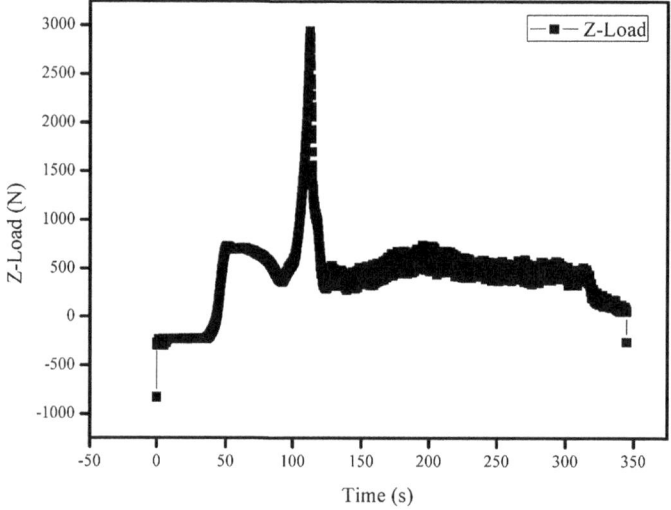

Fig. 6.14 Variation of axial load with time

nearly 5 s because the 5 s dwell time is one of the characteristics of the machine employed for the experiment. After this, the tool started travelling over the joint line and thus the force signal became nearly constant. At the end, as the tool retracted out of the joint, the force value approached to zero.

6.7.2.3 Tensile Strength

The tensile strength of the welded samples has been shown in Table 6.6. The tensile strength of the base material is 33 MPa. The maximum tensile strength obtained with the selected parameters is 14.63 MPa which is 44.34% of the base metal's strength. The effect of tool rotational speed on the tensile strength has been shown

Table 6.6 Tensile test results

S. No.	ω (rpm)	v (mm/min)	UTS (MPa)
1	500	10	13.48
2	500	20	7.02
3	500	30	4.5
4	600	10	14.23
5	600	20	9.67
6	600	30	6.3
7	800	10	14.63
8	800	20	13.85
9	800	30	10.7

UTS ultimate tensile strength

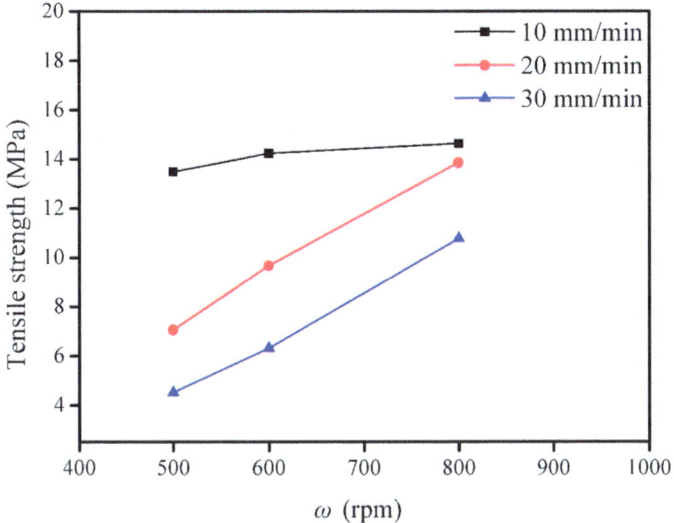

Fig. 6.15 Effect of tool rotational speed on tensile strength

in Fig. 6.15. It can be observed that the tensile strength is increasing with increase in the tool rotational speed. The frictional heat generation increases with the increase in the rotation of the tool which in turn increases the deformation rate of the base material. Thus, an optimum temperature has been achieved with the increasing tool rotation in this particular case, and hence, the tensile strength increases. At constant welding speed of 10 mm/min, the tensile strength obtained with 500 rpm is 13.48 MPa, and this value increases to 14.23 MPa with the increase in ω to 600 rpm. With further increase in ω to 800 rpm, the tensile strength again increases.

If we further go beyond $\omega = 800$ rpm, there will be probably no increase in the tensile strength. The reason to this is the material deformation rate, which also is increasing with the increase in the tool rotation rate. As discussed earlier, the thermoplastics are soft in nature and with high rotation of the tool shoulder portion over its surface, the material would char.

The effect of welding speed on the tensile strength has been shown in Fig. 6.16. It can be seen from the plot that the tensile strength decreases with the increase in the tool traverse speed. The time spent by a tool would decrease with high traverse speed and as such, the heat input in the process would decrease because of less friction between the tool and the work-piece. The tensile strength obtained with ω of 500 rpm and v of 10 mm/min is 13.48 MPa, and 7.02 MPa is the tensile strength obtained with ω of 500 rpm and 20 mm/min welding speed. The decrease in the tensile strength of the joint is probably because of the insufficient heat input which prevails due to the increase in the travel speed. With the further increase in the tool traverse speed, the tensile strength further decreases to 4.5 MPa (at ω of 500 rpm and v of 30 mm/min).

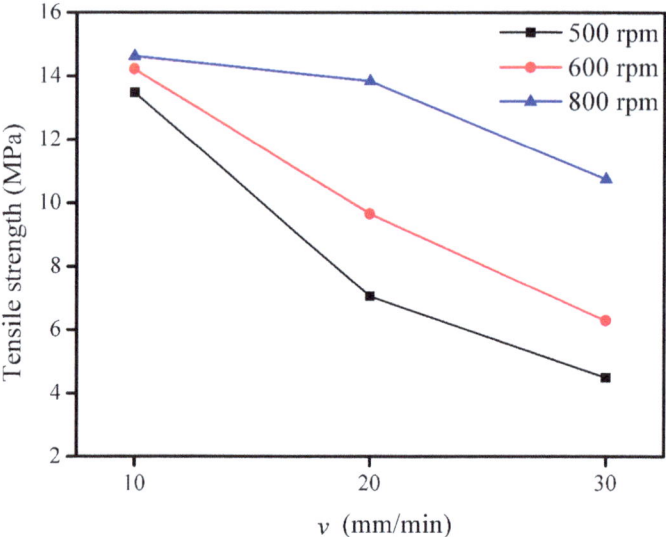

Fig. 6.16 Effect of welding speed on tensile strength

The tensile strength obtained with a ω of 800 rpm and a v of 30 mm/min is 10.7 MPa, while the tensile strength obtained with a ω of 500 rpm and a v of 30 mm/min is 4.5 MPa, and that of with 600 rpm and 30 mm/min is 6.3 MPa. The percentage change of tensile strength from 500 to 600 rpm is approximately 40%, while the change from 600 to 800 rpm is 69.9%. The increase in the tensile strength is because of the increase in ω and the irregularity in the result is probably because of the chosen step size of the parameters (ω) in this present case study. The lowest tensile strength obtained is 4.5 MPa employing the parameters: ω of 500 rpm and a v of 30 mm/min. With these parameters, the heat developed is the least since, the rotational rate is low and the speed is high. Thus, the friction is reduced and the tool spends less time, failing to achieve the required deformation and thus, the tool fails to homogenise the material resulting in low tensile strength. The highest tensile strength obtained in the present study is far away from that of the base metal's tensile strength. The possible reason to this is the traditional FSW tool designs, which are unable to properly mix the material and also are diminishing the strength of the work-piece with their rotational action on its surface.

6.7.2.4 Tensile Sample Fracture Observation

The base material, under the tensile test showed ductile nature with high amount of strain. But all the welded samples during the tensile test exhibited brittle failure nature. The fractured tensile specimen with highest tensile strength ($\omega = 800$ rpm, $v = 10$ mm/min) and the tensile specimen with lowest tensile strength

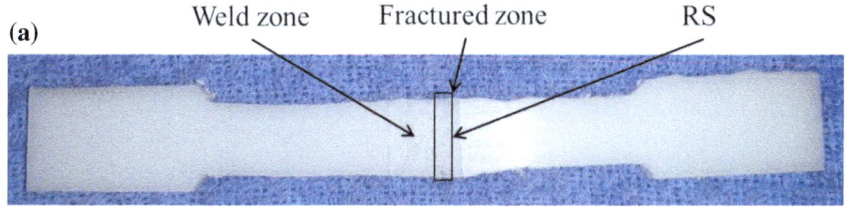

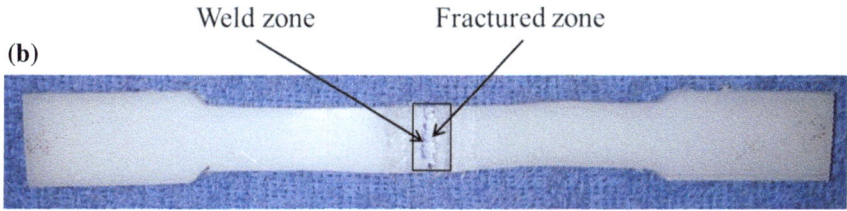

Fig. 6.17 Tensile fractured specimens. **a** $\omega = 800$ rpm; $v = 10$ mm/min, **b** $\omega = 500$ rpm; $v = 30$ mm/min

($\omega = 500$ rpm, $v = 30$ mm/min) have been shown in Fig. 6.17a, b, respectively. The tensile sample in (a) was fractured from the RS of the weld zone. Of course, the cavities in the sample were very few as compared to the other welded samples which reflect that the heat achieved was able to deform the work-pieces to a required level. The weld zone in (a) remain intact which show that the nugget was fully developed. The tensile specimen shown in (b) fractured from the weld zone. This shows the improper heat input with the chosen parametric combination and hence, the weld quality was very poor. The welded sample also had too many cavities which was an indication of the loss of material, and thus, improper material flow (discussed in Sect. 7.2.1).

6.7.2.5 Macro- and Micro-image Analysis

After the completion of the welding process, a small specimen was cut out from a welded sample and was cold mounted to visualise the weld zone. The mounted sample is shown in Fig. 6.18a and the macroscopic image of the same has been shown in Fig. 6.18b. This was particularly done because of the interesting view on the weld cross-section. The flow zone resembles a U-shape with the shoulder influenced zone higher than that of the pin influenced zone. The shoulder effect was limited on the surface of the work-piece since it can be observed that the weld flow zone decreases from the top to the bottom. The bottom of the work-piece is influenced by the pin. The small specimen after mounting was also visualised under bare eyes and it seemed as if there were empty spaces inside the weld zone. This

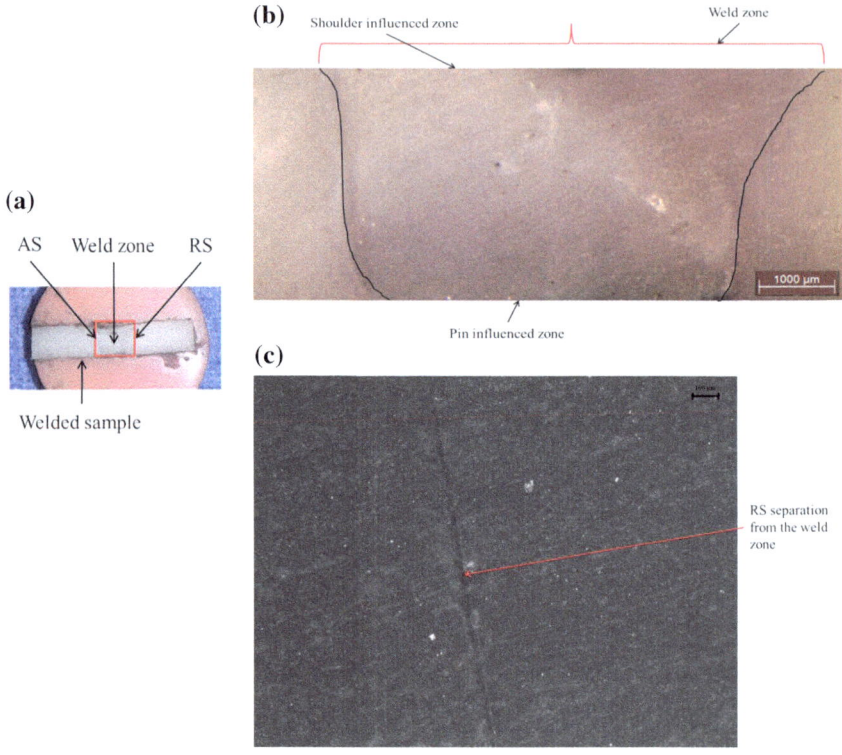

Fig. 6.18 **a** Cold mounted specimen of welded thermoplastics (ω = 800 rpm; v = 10 mm/min) and its, **b** macro-image, **c** micro-image

gave an indication that the materials are not getting properly mixed up, and hence, the homogenization has not been achieved. This could be a possible reason for the low tensile strength of the welded samples with respect to that of the base material. The sample was also viewed under the microscope to evaluate the weld zone. It can be observed from Fig. 6.18c that a line is present that is acting like a boundary between two zones. The left to that line is the weld zone and the line is the RS. This result proved that the traditional FSW tool is unable to make the material flow through the edges well.

6.8 Conclusion

The engineering thermoplastics have the potential to replace the metallic components in various application areas by reducing the weight and improving the efficiency. In order to practically employ the thermoplastics in industrial equipment and structures, efficient methods have to be developed to join them. The chapter focuses on the joining aspects of polymeric materials through FSW. The nature of the polymers has been described and the various available methods, such as, mechanical fastening, adhesive bonding and welding techniques to join them have been outlined. The methodologies of the aforementioned techniques, merits and demerits, applications, etc. have also been discussed. Adhesive bonded joints are not reliable and the mechanical fastening methods introduce unwanted stresses in the polymeric components. The fusion welding techniques, with high heat application could destroy the soft nature of the polymer and as such, a solid state joining technique would satisfy the need. A detailed idea of the FSW process has been presented to highlight the increasing demand and versatility of the welding technique. At last, a case study has been presented where a polymer, HDPE has been welded employing three values of ω (500, 600 and 800 rpm) and three values of v (10, 20 and 30 mm/min). A 1° tilt angle was applied during every experiment. Post weld analyses such as the weld visual test and tensile tests have been performed. The following conclusions have been drawn:

- The welding of thermoplastic materials is a gruelling task since they have low melting point and are delicate materials as compared to the metals.
- The thermoplastics have extremely low thermal conductivity; hence make the welding heat concentrated at a same location. Also, they have high value of coefficient of thermal expansion.
- Traditional FSW tools are not suitable for welding the polymeric materials, since the shoulder action is creating adverse effects on the material by disrupting its inner fibrous molecules. The use of stationary shoulder and double shoulder with fully inserted pin are likely to be the tools for polymers.
- In the present case study, the employed parameters have successfully welded the HDPE sheets with good finish and negligible defects, but the achieved joint strength efficiency is not much appreciable.
- The tensile strength in the study has been found to be increasing with the increase in ω and decrease with increase in the v. The maximum strength achieved is 14.63 MPa employing welding parameters of 800 rpm, 10 mm/min and 1°.

References

1. Chawla, K.K.: Composite Materials. Springer, New York (1998)
2. Norris, G., Wagner, M.: Boeing 787 Dreamliner (2009). Available: http://www.modernairliners.com/boeing-787-dreamliner/boeing-787-dreamliner-specs/
3. Plastics—The Facts 2010, An analysis of European plastics production, demand and waste data (2010)
4. Fried, J.R.: Polymer Science and Technology, 3rd edn., vol. 40, no. 6. Prentice Hall, Englewood Cliffs (2014)
5. Amancio Filho, S.T.: Friction riveting development and analysis of a new joining technique for polymer-metal multi-material structures (2011)

54. Mendes, N., Loureiro, A., Martins, C., Neto, P., Pires, J.N.: Effect of friction stir welding parameters on morphology and strength of acrylonitrile butadiene styrene plate welds. Mater. Des. **58**, 457–464 (2014)
55. Mendes, N., Loureiro, A., Martins, C., Neto, P., Pires, J.N.: Morphology and strength of acrylonitrile butadiene styrene welds performed by robotic friction stir welding. Mater. Des. **64**, 81–90 (2014)
56. Arici, A., Sinmaz, T.: Effects of double passes of the tool on friction stir welding of polyethylene. J. Mater. Sci. **40**(12), 3313–3316 (2005)
57. Aydin, M.: Effects of welding parameters and pre-heating on the friction stir welding of UHMW-polyethylene. Polym. Plast. Technol. Eng. **49**(6), 595–601 (2010)
58. Bozkurt, Y.: The optimization of friction stir welding process parameters to achieve maximum tensile strength in polyethylene sheets. Mater. Des. **35**, 440–445 (2012)
59. Payganeh, G.H., Arab, N.B.M., Asl, Y.D., Ghasemi, F.A., Boroujeni, M.S.: Effects of friction stir welding process parameters on appearance and strength of polypropylene composite welds. Int. J. Phys. Sci. **6**(19), 4595–4601 (2011)
60. Jaiganesh, V., Maruthu, B., Gopinath, E.: Optimization of process parameters on friction stir welding of high density polypropylene plate. Procedia Eng. **97**, 1957–1965 (2014)
61. Melendez, M., Tang, W., Schmidt, C., McClure, J.C., Nunes, A.C., Murr, L.E.: Tool forces developed during friction stir welding, pp. 1–38 (2013)
62. Squeo, E.A., Bruno, G., Guglielmotti, A., Quadrini, F.: Friction stir welding of polyethylene sheets. Ann. "DUNĂREA JOS" Univ. Galati Fascicle V Technol. Mach. Build. 241–146 (2009)
63. Bagheri, A., Azdast, T., Doniavi, A.: An experimental study on mechanical properties of friction stir welded ABS sheets. Mater. Des. **43**, 402–409 (2013)
64. Shazly, M., El-raey, M.: Friction stir welding of polycarbonate sheets, pp. 555–563 (2014)
65. Saeedy, S., Givi, M.K.B.: Proc. Inst. Mech. Eng. Part B J. Eng. Manuf. Investig. Eff. Crit. Process Parameters (2011)
66. Amancio-Filho, S.T., dos Santos, J.F.: Joining of polymers and polymer-metal hybrid structures: Recent developments and trends. Polym. Eng. Sci. **49**(8), 1461–1476 (2009)

Chapter 7
SiC and Al$_2$O$_3$ Reinforced Friction Stir Welded Joint of Aluminium Alloy 6061

Md. Aleem Pasha, P. Ravinder Reddy, P. Laxminarayana
and Ishtiaq Ahmed Khan

Abstract This research presents the enhancement of mechanical properties of friction stir welded portion of aluminium alloy 6061 by incorporating additional reinforcing particulates of silicon carbide and aluminium oxide at weld interface. Friction stir welding (FSW) of AA6061, each plate of 200 mm × 100 mm × 4 mm thickness with silicon carbide and aluminium oxide as reinforcement at weld interface in four different volume proportions and without reinforcement are performed on vertical milling machine. In the present research, comparison has been done between mechanical properties of silicon carbide and aluminium oxide reinforced welded joints. Silicon carbide and aluminium oxide have been added as reinforcement by creating separate geometry, at the edges, where the welding is interfaced with four different volume proportions such as 10, 15, 25 and 30%. Tool steel of H13 grade is used as friction stir welding tool. The tool has outer shoulder diameter of 18 mm, pin diameter at the top of 7 mm and 5 mm at the bottom, and pin length of 3.7 mm. A rotational speed of 1120 rpm and transverse speed of 40 mm/min were selected. Quality assessment is carried out by visual inspection and non-destructive testing using fluorescent and radiography to reveal the surface and volumetric defects. Mechanical testings including tensile test, impact test, bend test and hardness test were conducted to study the behaviour of reinforced and unreinforced friction stir welded joints. Metallurgical evaluation has been performed by capturing the microstructures of base materials and at different zones of

Md. Aleem Pasha (✉)
Mechanical Engineering Department, CBIT, Hyderabad, India
e-mail: aleemphd2013@gmail.com

P. Ravinder Reddy
Mechanical Engineering Department and Principal, CBIT, Hyderabad, India
e-mail: reddy.prr@gmail.com

P. Laxminarayana
Mechanical Engineering Department, Osmania University, Hyderabad, India
e-mail: laxp@rediffmail.com

I. A. Khan
TATA Technologies, Pune, India
e-mail: khan12341@gmail.com

© Springer Nature Singapore Pte Ltd. 2019
U. S. Dixit and R. G. Narayanan (eds.), *Strengthening and Joining by Plastic Deformation*, Lecture Notes on Multidisciplinary Industrial Engineering,
https://doi.org/10.1007/978-981-13-0378-4_7

nugget, heat-affected zone (HAZ) by optical microscope to reveal the grain size and grain refinement at different zones. The experimental results indicate that the reinforcing particulates and percentage of reinforcing particulates have a major influence on the mechanical properties of friction stir welded joint. From microstructures, it has been shown that the addition of SiC and Al_2O particles decreased the grain size and increased the strength of the joints.

7.1 Introduction

Friction Stir Welding (FSW) is a relatively new solid-state metal joining process (without melting) which first found its applications in niche markets such as joining of aerospace aluminium alloys, besides dissimilar welds, which are difficult to weld with traditional welding techniques. Joining of large panels, which cannot be easily heat-treated post-weld to recover temper characteristics, is another common interest for FSW to be preferred for industrial applications besides its other benefits. It was invented and experimentally proven by Wayne Thomas and his colleagues at The Welding Institute, Cambridge, UK in December 1991, and then TWI filed for worldwide patent protection in December of that year.

At its invention in 1991, there were many reasons to believe that FSW was a major breakthrough in the area of metal joining. FSW was able to reach a joint strength beyond conventional fusion welding processes while avoiding typical drawbacks like intensive light, fumes, spatter, high currents, shielding gas and consumables.

FSW uses a non-consumable rotating cylindrical tool having a shoulder and a pin (Fig. 7.1). The shoulder is pressed against the surface of the materials being welded, while the probe is forced between the two components by a downward force. The combined effect of tool rotation and translation involves heat generation by friction between tool and plates being welded, and severe plastic deformation of the workpiece material. This softens the material being welded. Then, the material is transported from the front of the tool to the trailing edge where it is forged into a joint. Probe is the part of the tool which is inserted into the workpiece by axial force which shears the material in front of the tool and moves the same behind the tool to make the joint [1, 2].

Elangovan et al. [3] attempted to study the effect of different tool pin profiles such as straight cylindrical, tapered cylindrical, threaded cylindrical, square and triangular. Tool with square pin profile has been given good results compared to other pin profiles irrespective of shoulder diameter of the tools. Tool shoulder diameters such as 15, 18 and 21 mm were used in the formation of friction stir processing zone in AA6061. From investigation, the important conclusions are derived that the tool with 18 mm shoulder diameter produced defect-free FSP

Fig. 7.1 Friction stir welding
process

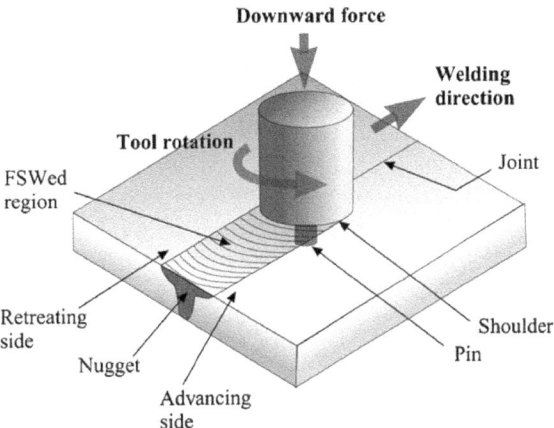

region, irrespective of tool pin profiles. The joint fabricated using square pin pro-filed tool with shoulder diameter of 18 mm showed superior tensile properties.

Karthikeyan and Mahadevan [4] have been conducted research on FSW of AA6351 alloy by adding SiC particulates in the weld zone. Results proved that the mechanical properties of welded portion are enhanced by 33%, due to the restriction of grain boundaries growth by pinning of added particles.

Sun and Fujii [5] conducted research on FSW of pure copper with SiC particles of size 5 μm which were inserted between the two adjacent plates of 1 mm gap. It was observed that, during FSW process, SiC particles simulated nucleation in the dynamic recrystallization of Cu. As a result, grain structure was refined and the hardness increased.

From the literature, it is observed that the mechanical properties of friction stir welded joints were less than the mechanical properties of base material. In the present research, silicon carbide and aluminium oxide particles were added to weld stir zone of FSW of AA 6061 to enhance the metallurgical and mechanical prop-erties. Metal matrix composite was formed at weld portion due to the reinforcement of additional SiC and Al$_2$O$_3$ ceramic particles. Reinforcing materials of silicon carbide and aluminium oxide were added at weld interface with four volume proportions of 10, 15, 25 and 30%. The effect of volume proportions of reinforcing materials on microstructure and mechanical properties of friction stir welded joints of AA6061 was investigated. Microstructure and mechanical properties of the SiC and Al$_2$O$_3$ reinforced friction stir welds of AA6061 are differentiated with those of the unreinforced FSW samples.

7.2 Experimentation of Reinforced and Unreinforced Friction Stir Welding Process

The material used for friction stir welded butt joints with reinforcement was 4-mm-thickness aluminium alloy 6061 plate. The plate of AA6061 was machined to required dimensions of 200 mm × 100 mm by using shearing machine. The plates were machined on vertical milling machine at the edges of the plates in stepped shape, to form a groove after placing their stepped ends in contact to fill the reinforced particles for different cases. Before welding, the plates were cleaned by using ethanol chemical to remove surface contaminations. The plates were fixed on vertical milling machine of HMT FM-2 having capacity of 10 H.P, 3000 rpm, for friction stir welding as shown in Fig. 7.2.

The Al_2O_3 powder and SiC powder were filled into a groove individually, as per different cases considered on the plate of 4 mm thickness before FSW was carried out. H13 tool steel, which is shown in Fig. 7.3, was used as welding tool. The above-mentioned vertical milling machine was used to perform the welding process. The tool is placed in the tool holder of collet 18 mm diameter. The initial welding has been done on the workpieces of 200 mm × 100 mm aluminium alloy

Fig. 7.2 Vertical milling machine used for FSW

Fig. 7.3 FSW tool used for experiments

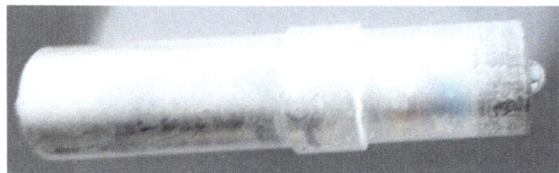

6061 plates without reinforcement (Plates are unstepped machined) by placing them on vertical milling machine which gets operated at rotational speed of 1120 rpm and feed rate of 40 mm/min.

In the second section, four friction stir welds were performed on the AA6061 stepped end plates by filling the silicon carbide (SiC) particulates as reinforcement with four volume proportions of 10, 15, 25 and 30% in the formed groove of plates after aligning their stepped faces in the fixed condition on the vertical milling machine which operates at rotational speed of 1120 rpm and feed rate of 40 mm/min.

In third section, four friction stir welds were performed to the AA6061 stepped end plates by filling the aluminium oxide (Al_2O_3) particulates as reinforcement with four volume proportions of 10, 15, 25 and 30% in the formed groove of plates after aligning their stepped faces in the fixed condition on the vertical milling machine which operates at rotational speed of 1120 rpm and feed rate of 40 mm/min.

Finally, the welded plates were placed under different quality inspections of non-destructive and destructive tests. Radiography test and fluorescent liquid penetrate tests were performed under non-destructive tests. Destructive tests like tensile test, bend test, impact test and Rockwell hardness tests were performed. Metallurgical characterization was performed using optical microstructures.

7.2.1 Friction Stir Welding Tool and Tool Material

The tool material used in the experiments was tool steel grade of H13. The designed and the prepared tool is shown in Fig. 7.3. The chemical composition of tool is indicated in Table 7.1.

7.2.2 Workpiece Materials

Commercial aluminium alloy 6061 sheet of 200 mm × 100 mm × 4 mm was used as workpiece material as shown in Fig. 7.4. The chemical composition of workpiece material is indicated in Table 7.2.

Table 7.1 Chemical composition of H13 tool steel

Element	C	Mn	Cr	Mo	V	Si	Fe
Weight (%)	0.40	0.35	5.20	1.30	0.95	1.0	90.8

Fig. 7.4 AA6061 plates used
for FSW

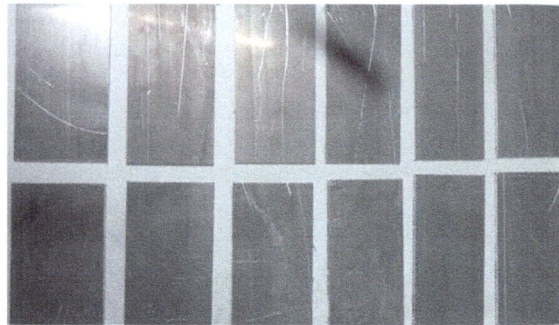

Table 7.2 Chemical composition of AA6061

Element	Mg	Si	Fe	Cu	Cr	Mn	Zn	Ti	Al
Weight (%)	0.9	0.62	0.33	0.28	0.17	0.06	0.02	0.02	97.6

7.2.3 Reinforced Particulates

In the present research work, silicon carbide particles of 400 mesh (37 µm) as
shown in Fig. 7.5a were used as reinforcement. Silicon carbide is composed of
tetrahedral of carbon and silicon atoms with strong bonds in the crystal lattice.
Silicon carbide of thermal conductivity 120 W/m °K, specific heat 750 J/kg °K,
density 3.1×10^{-2} tones/mm^3, thermal expansion 4×10^{-6}/°C and Young's
modulus 410×10^3 N/mm^2 was used.

Aluminium oxide of finer ceramic powder as shown in Fig. 7.5b was used as
reinforced particulates at weld interface. Commercially available aluminium oxide
of finer powder of thermal conductivity 18 W/m °K, specific heat 880 J/kg °K,
density 3.69×10^{-2} tones/mm^3, thermal expansion 8.1×10^{-6}/°C, and Young's
modulus 300×10^3 N/mm^2 was added as reinforcement.

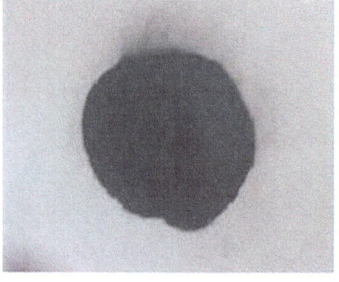

(a) Silicon Carbide   (b) Aluminium oxide

Fig. 7.5 Reinforcement particulates

Table 7.3 Process
parameters for FSW

Process parameters	
Rotation speed (rpm)	1120
Welding speed (mm/min)	40
Tool shoulder diameter (mm)	18
Pin diameter at top (mm)	7
Pin diameter at bottom (mm)	5
Pin length (mm)	3.7
Tilt angle (°)	0
Pin profile	Conical

7.2.4 Process Parameters in FSW

In the present work, all the process parameters used for FSW as shown in Table 7.3 were maintained as constant for all the four different volume proportions of reinforcements.

7.2.5 Geometry of Groove at Ends of Plates to Incorporate Reinforcement

To add the reinforcement particulates, the workpiece ends were machined on a milling machine in the form of a step. The groove is formed at interface of two-stepped ends of two plates producing a butt joint. Width and depth of groove were decided according to volume percentage variation of reinforcement particulates. The buttface of the job to be welded after filling reinforcement is shown in Fig. 7.6.

Silicon carbide and aluminium oxide have been added as reinforcement by creating separate geometry, at the edges where the welding is interfaced with four different volume proportions such as 10, 15, 25 and 30%. Volume proportions are varied by changing groove size like width and depth of groove as shown in Table 7.4. In case I, II, III and IV, 10, 15, 25 and 30% of reinforce particulates were added at weld interface. Width and depth of the groove in each plate is varied as shown in Table 7.4 to get required volume proportion.

7.2.6 FSW Experiments

Aluminium alloy 6061 reinforced with silicon carbide and aluminium oxide at different volume fractions of 10, 15, 25 and 30% and without reinforcement were successfully friction stir welded on vertical milling machine at 1120 rpm rotational

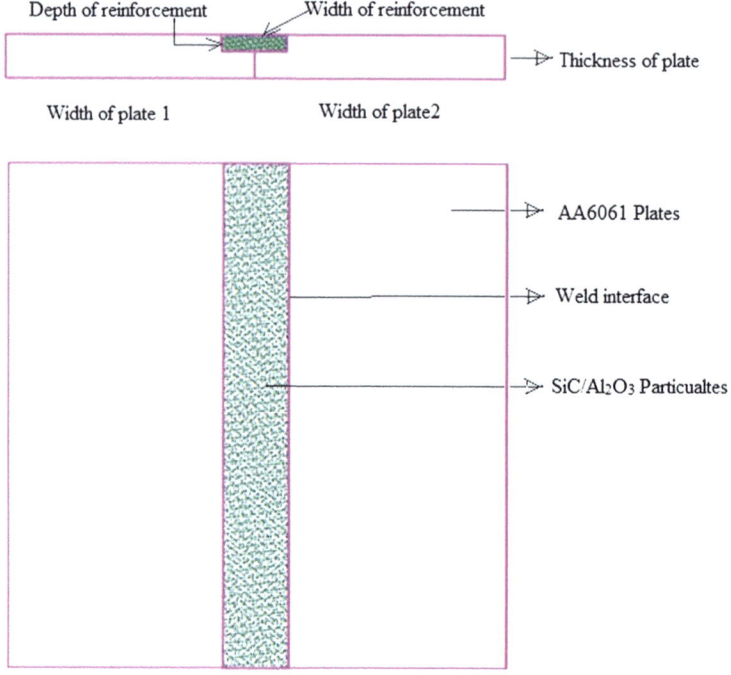

Fig. 7.6 Schematic representation of workpiece with reinforcement groove

Table 7.4 Different groove sizes for different volume proportions of reinforcement

Case	% of reinforcement	Size of the groove for reinforcement	
		Width (mm)	Depth (mm)
I	10	0.5	2
II	15	0.5	3
III	25	1	2.5
IV	30	1	3

speed and 40 mm/min transverse feed. Figure 7.7 shows various weld specimens with different reinforcements and volume proportions of reinforcements.

7.2.7 Non-destructive Testing of FSW Sheets

Non-destructive tests of fluorescent liquid penetrate test as per ASTM E165 and X-ray radiography tests as per ASME Sec-IX-2013 QW-191 were conducted on all the specimens as shown in Fig. 7.8, after FSW to evaluate weld quality. It was found that all the specimens were defect-free.

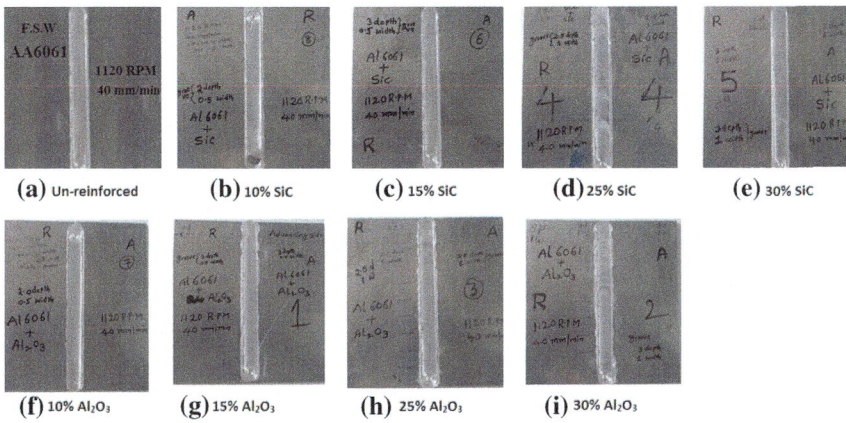

Fig. 7.7 Samples as per volume % and reinforcement type

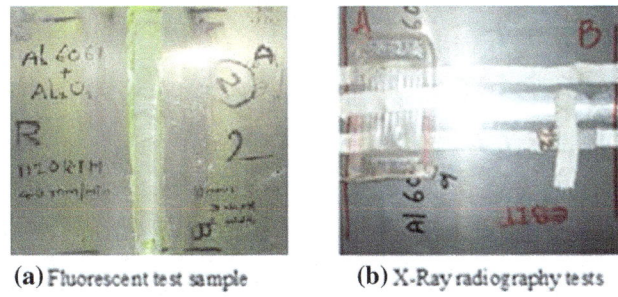

Fig. 7.8 Non-destructive test samples

7.2.8 Destructive Testing of FSW Joints

After FSW, the welded plates were tested to evaluate the mechanical behaviour of welded joints such as ultimate tensile strength, yield strength, percentage of elongation, bending strength, impact strength and hardness as shown in Fig. 7.9.

7.3 Results and Discussion

The results of mechanical and metallurgical behaviour of SiC and Al₂O₃ reinforced and unreinforced friction stir welded joints of aluminium alloy 6061 are tabulated and discussed. The effect of SiC and Al₂O₃ as reinforcement at weld interface on the mechanical properties of welded portion is discussed. The effect of volume proportions of reinforcement particulates over the mechanical properties of welded

(a)

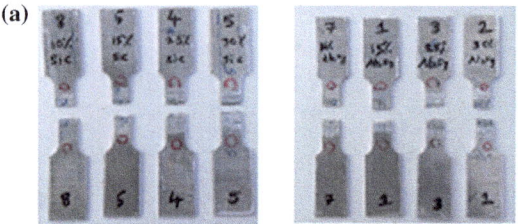

Tensile specimens of FSW AA 6061 after tensile test

(b)

Tensile specimen of AA 6061 (Base metal) after tensile test

(c)

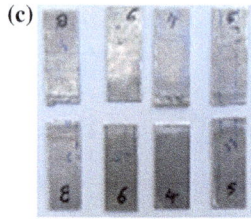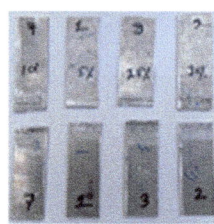

FSW specimens of AA6061 after bend test

(d)   **(e)**

Bend specimen of AA 6061   Bend specimen Unreinforced FSW

(f)

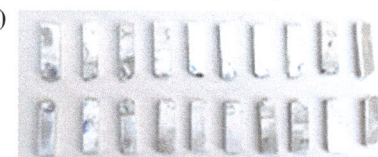

Impact test tested specimens of FSW AA6061

(g)

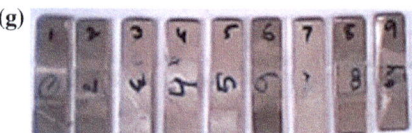

FSW specimens of AA6061 for hardness test

Fig. 7.9 Different destructive test samples

portion is also discussed and further addresses the metallurgical behaviour of reinforced and unreinforced friction stir welded joints of aluminium alloy 6061.

7.3.1 Test Results of SiC Reinforced FSW Joints

The mechanical test results of SiC reinforced and unreinforced welded specimens are given in Table 7.5 for comparison. The addition of SiC particulates in the weld region had improved the mechanical properties like tensile strength, yield strength and hardness as compared to unreinforced welded joints. This is due to the existence of second-phase particles causing in the creation of dislocation loops and dislocation cell structure around the particles during deformation. As a result, the grain size decreases with increase in strength and hardness [6]. The added SiC particles in the weld region had improved the mechanical properties in the weld zone on the account of refinement of grains, uniform mixing of the parent material and the SiC ceramic particles in that region. Hence, the SiC particles play a vital role in controlling the effect of grain coarsening and softening of the material. The addition of SiC particulates in the weld zone lowers the impact strength, elongation and bending quality because of brittleness of the added SiC particulates.

With increase in volume proportions of SiC particulate from 10 to 25%, the yield strength, tensile strength and hardness were found to increase. This is believed to be because of increase in the heat input and decrease in the grain size at weld interface with increase in the volume proportions of SiC particulates at weld interface. Grain size is changed from 11–12 to 3–4 μm at 10–25% volume proportion.

The tensile strength, yield strength and hardness were lowered at 30% volume proportion of SiC particulates. This is because of less heat developed at weld region resulting in the lack of proper stirring and non-homogenous distribution of SiC particles. The spacing between interparticles is more and hence the grain size increase. This decreases the strength properties [6, 7].

Table 7.5 Mechanical test results of FSW joints of AA6061

Mechanical properties	Base material	Unreinforced FSW joint	Volume proportion of SiC			
			10%	15%	25%	30%
Tensile strength (MPa)	315.26	163.81	221.15	252.00	303.05	263.28
Yield strength (MPa)	265.26	138.15	191.18	217.42	263.68	261.03
% elongation	17.240	9.00	7.47	7.02	6.09	3.80
Impact strength (J)	9	12	8	8	6	6
Hardness (BHN)	55	68	93	96	104	101
Bend quality	Not failed	Not failed	Failed	Failed	Failed	Failed

Table 7.6 Efficiencies of unreinforced and SiC reinforced FSW joints over base material

Mechanical properties	Unreinforced FSW joint	Volume proportion of SiC			
		10%	15%	25%	30%
Tensile strength (MPa)	51.97	70.15	79.94	96.13	83.51
Yield strength (MPa)	52.09	72.08	81.97	99.41	98.40
% elongation	52.21	43.40	40.72	35.33	22.05
Impact strength (J)	133.33	88.89	88.89	66.67	66.67
Hardness (BHN)	123.63	169.09	174.54	189.09	183.63
Bend quality	Not failed	Failed	Failed	Failed	Failed

The efficiencies of the reinforced and unreinforced FSW joints were evaluated with respect to base material properties by assuming base material properties efficiencies as 100%. It is observed from Table 7.6 that the tensile strength efficiency of unreinforced friction stir welded joint is 51.97%. The tensile strength efficiency of weld has enhanced by the addition of SiC particulates. This is due to the refinement of grain size at welded portion. Maximum tensile strength efficiency was obtained at 25% volume proportions of SiC. The efficiency of elongation is more in unreinforced welded joints than in reinforced welded joints. Lowest elongation efficiency is obtained at 30% volume proportions of SiC reinforcement. The loss of elongation efficiency could be due to the presence of ceramic particulates which are brittle in nature. Impact strength efficiency of SiC reinforced friction stir welded joints is lower than the base material and unreinforced welded joints. Hardness efficiency of unreinforced and reinforced friction stir welded joint of AA6061 was more than the base material. By the addition of SiC particulates at weld interface, the hardness efficiency of joint was improved compared to that of hardness efficiency of unreinforced friction stir welded joint. The maximum hardness efficiency was obtained at 30% volume proportions of SiC reinforcement. The bend strength efficiency of unreinforced welded joint was same as that of base material bend strength efficiency. By the addition of SiC particulates at welded portion, the welded joint was failed without bend. This is due to brittle and hard nature of SiC ceramic particulates.

7.3.2 Mechanical Test Results of Al₂O₃ Reinforced FSW Joints

7.3.2 *Mechanical Test Results of Al_2O_3 Reinforced FSW Joints*

The mechanical test results of Al_2O_3 reinforced and unreinforced welded specimens are given in Table 7.7 for comparison. The addition of Al_2O_3 particulates in the weld region had improved the mechanical properties like tensile strength, yield strength and hardness as compared to unreinforced welded joints. This is due to the existence of second-phase particles causing the formation of dislocation loops and

Table 7.7 Mechanical test results of base material AA6061, unreinforced and Al$_2$O$_3$ reinforced FSW joints

Mechanical properties	Base material	Unreinforced FSW joint	Al$_2$O$_3$ volume proportions			
			10%	15%	25%	30%
Tensile strength (MPa)	315.269	163.815	187.98	197.36	225.57	236.85
Yield strength (MPa)	265.260	138.157	177.79	182.89	206.71	210.52
% elongation	17.240	9.00	8.06	8.26	8.6	9
Impact strength (J)	9	12	9.6	9.0	8.2	8.0
Hardness (BHN)	55	68	80	84	88	91
Bend quality	Not failed	Not failed	Failed	Failed	Failed	Failed

dislocation cell structure around the particles during deformation. As a result, the grain size decreases with increase in strength and hardness. The added Al$_2$O$_3$ particles in the weld region had improved the mechanical properties in the weld zone on the account of refinement of grains, uniform mixing of the parent material and the Al$_2$O$_3$ ceramic particles in that region; hence, the Al$_2$O$_3$ particles play a vital role in controlling the effect of grain coarsening and softening of the material. The addition of Al$_2$O$_3$ particulates in the weld zone lowers impact strength, elongation and bend quality, because of brittleness of the added Al$_2$O$_3$ particulates.

With increase in volume proportions of Al$_2$O$_3$ particulate from 10 to 30%, the yield strength, tensile strength and hardness were found to increase. This is believed to be because of increase in the heat input and decrease in the grain size at weld interface with increase in the volume proportions of Al$_2$O$_3$ particulates at weld interface.

The efficiencies of the reinforced and unreinforced FSW joints were evaluated with respect to base material properties by assuming base material properties efficiencies as 100%. It is observed from Table 7.8 that the tensile strength efficiency of unreinforced friction stir welded joint is 51.97%. The tensile strength and hardness efficiency of weld have enhanced by the addition of Al$_2$O$_3$ particulates.

Table 7.8 Efficiencies of unreinforced and Al$_2$O$_3$ reinforced FSW joints of AA6061 over base material

Mechanical properties	Unreinforced FSW joint	Al$_2$O$_3$ volume proportions			
		10%	15%	25%	30%
Tensile strength (MPa)	51.97	59.63	62.61	71.55	75.13
Yield strength (MPa)	52.09	67.03	68.95	77.93	79.37
% elongation	52.21	46.76	47.92	49.89	52.21
Impact strength (J)	133.33	106.6	100	91.12	88.89
Hardness (BHN)	123.63	145.45	152.72	160	165.45
Bend quality	Not failed	Failed	Failed	Failed	Failed

This is due to refinement of grain size at welded portion. Maximum tensile strength and hardness efficiency were obtained at 30% volume proportions of Al_2O_3. The efficiency of elongation is more in unreinforced welded joints compared to 10, 15 and 25% reinforce welded joints. Lowest elongation efficiency is obtained at 10% volume proportions of Al_2O_3 reinforcement. Elongation efficiency of 30% reinforce welded joints is equal to unreinforced welded joints. Impact strength efficiency of Al_2O_3 reinforced friction stir welded joints is lower than the base material and unreinforced welded joints. The bend strength efficiency of unreinforced welded joint was same as that of base material bend strength efficiency. By addition of Al_2O_3 particulates at welded portion, the welded joint was failed without bend. This is due to brittle and hard nature of Al_2O_3 ceramic particulates.

7.3.3 Comparison of Addition of SiC and Al_2O_3 Particles

A comparative assessment of tensile strength and yield strength of SiC reinforced and Al_2O_3 reinforced FSW joints has been done in Fig. 7.10. The unreinforced joints perform worse than base material and reinforced joints perform in between. Moreover, the performance of SiC reinforced joints is better than Al_2O_3 reinforced joints in all the volume proportions. This is believed due to the better strength and mechanical properties of SiC particulates as compared to Al_2O_3 particulates.

Figure 7.11 reveals that the hardness of 10, 15, 25 and 30% of SiC reinforcement is 69.09, 74.54, 89.09 and 83.63% more than that of parent material hardness, respectively. The hardness of 10, 15, 25 and 30% of Al_2O_3 reinforcement was 45.45, 52.72, 60 and 65.45% more than that of parent material hardness. There is about 20–29% enhancement in hardness when SiC is used as reinforcement.

Figure 7.12 reveals that the impact strength of 10, 15, 25 and 30% of SiC reinforcement is 88.89, 88.89, 66.67 and 66.67% of parent material impact strength,

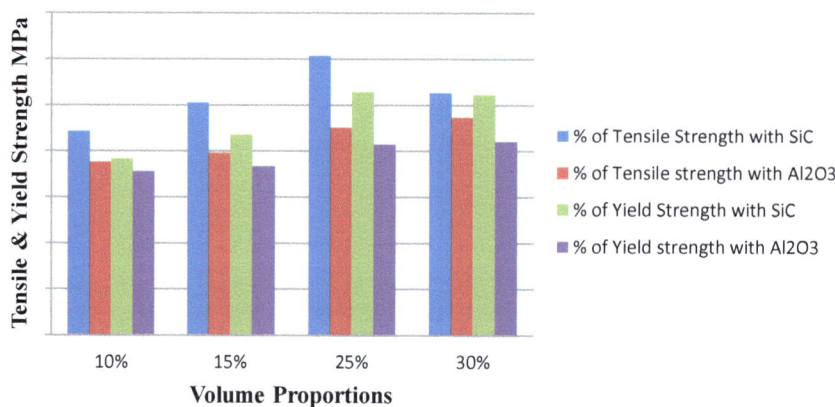

Fig. 7.10 Comparison of tensile and yield strength between SiC and Al_2O_3 reinforced FSW joints

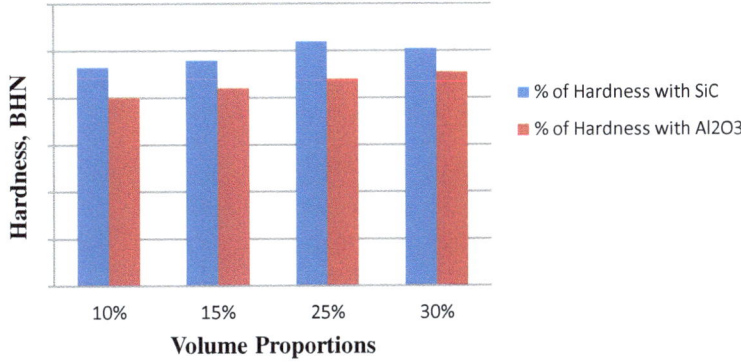

Fig. 7.11 Comparison of hardness between SiC and Al$_2$O$_3$ reinforced FSW joints

respectively. The impact strength of 10, 15, 25 and 30% of Al$_2$O$_3$ reinforcement is 106.6, 100, 91.12 and 88.89% of parent material impact strength. There is about 22.22–66.67% deterioration in impact strength when SiC is used as reinforcement. Figure 7.12 also reveals that the percentage of elongation of 10, 15, 25 and 30% of SiC reinforcement is 43.40, 40.72, 35.33 and 22.05% of parent material percentage of elongation, respectively. The percentage of elongation of 10, 15, 25 and 30% of Al$_2$O$_3$ reinforcement is 46.76, 47.92, 49.89 and 52.21% of parent material percentage of elongation, respectively. The performance of Al$_2$O$_3$ reinforced joints is better than SiC reinforced joints in all the volume proportions.

Table 7.9 reveals that the mechanical properties of tensile strength, yield strength and hardness are enhanced more in the case of SiC reinforcement as compared to Al$_2$O$_3$ reinforcement. The elongation and impact strength are enhanced more in the case of Al$_2$O$_3$ reinforcement than with SiC reinforcement. Bend quality remains same in both the cases. This is believed due to the brittleness and larger strength of SiC particles as compared to Al$_2$O$_3$ particulates.

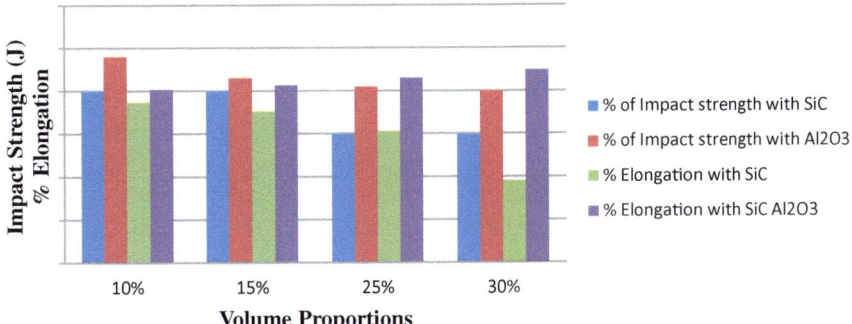

Fig. 7.12 Comparison of impact strength and elongation between SiC and Al$_2$O$_3$ reinforced FSW joints

Table 7.9 Percentage enhancement and deterioration of mechanical properties of SiC reinforced joints over Al_2O_3 reinforced joints

Mechanical properties	% enhancement and deterioration of mechanical properties			
	Volume proportion of reinforcements			
	10%	15%	25%	30%
Tensile strength (MPa)	10.52 (E)	17.33 (E)	24.58 (E)	8.38 (E)
Yield strength (MPa)	5.05 (E)	13.02 (E)	21.48 (E)	19.03 (E)
% elongation	3.36 (D)	7.20 (D)	14.56 (D)	30.16 (D)
Impact strength (J)	17.71 (D)	11.11 (D)	24.45 (D)	22.22 (D)
Hardness (BHN)	23.64 (E)	21.82 (E)	29.09 (E)	18.18 (E)
Bend quality	Failed	Failed	Failed	Failed

E enhanced; *D* deteriorated

7.3.4 Microstructural Analysis

Metallurgical testing has been done on reinforced and unreinforced FSW joints by using a computerized metallurgical microscope. Microstructural observations were carried out to capture microstructures and evaluate grain size at different zones of FSW joint using a digital image computerized optical microscope of model LABEX FE PRO 900 with 100× to 800× magnification as shown in Fig. 7.13. Dino-Eye C-Mount Camera of 10× magnification was mounted on optical microscope. The microstructures were examined to evaluate grain size by using Met Image Software [8].

After successful joining, specimens of size 15 mm × 10 mm × 4 mm from both the sides of weld and from centre of the weld normal to the weld line were cut using a wire cut electric discharge machine to study the metallurgical behaviour at the stir zone (nugget), heat-affected zone (HAZ) and parent metal. The specimens were then polished using different graded emery sheets followed by polishing using diamond paste on disc polishing machine with velvet cloth and cleaned with ethanol. Chemical etching of the polished specimens was done using 0.5% hydrofluoric acid for 10 s, rinsed distilled water and dried in hot air.

The microstructure of aluminium alloy 6061 shows that the base material contains longer elongated fine grains as shown in Fig. 7.13a. The average grain size was evaluated to be 20–22 μm using Met Image Software of version 4.11. Microstructure of unreinforced FSW joint of aluminium alloy 6061 consists of fine, equiaxed and recrystallized grains of average size of 14–18 μm at weld nugget as shown in Fig. 7.13b. After FSW, the grain size has become finer at the weld zone than in the base material. The microstructure of HAZ of unreinforced and SiC and Al_2O_3 reinforced FSW joints were investigated and the results revealed that the HAZ consists of grains having approximately the same size as that of base metal. This is because of the fact that the HAZ is only exposed to the welding heat, but not the deformation and recrystallization. The average grain size of reinforced and unreinforced FSW in the HAZ has been observed to be 20 μm (Fig. 7.13c). Since

Fig. 7.13 Microstructures of base material and reinforced and unreinforced FSW joints of AA6061

there is no change in the microstructures of HAZ of reinforced and unreinforced welds, only one microstructure is shown (Fig. 7.13c). The microstructures of nugget zone of unreinforced and SiC and Al₂O₃ FSW joints were investigated and results revealed that the nugget zone consists of fine, equiaxed and recrystallized grains as shown in Fig. 7.13. In reinforced FSW joints, the grain size has become finer at the weld zone than in the base material and unreinforced FSW joint.

Table 7.10 Grain sizes of AA6061 and reinforced and unreinforced FSW joints of AA6061

Cases	Grain size in µm		
		SiC	Al_2O_3
Base material	20–22	–	–
Unreinforced	14–18	–	–
10%	–	11–12	13.5
15%	–	8–9	11
25%	–	3–4	9.5
30%	–	5–6	7

The grain size of unreinforced FSW joints of AA6061 is decreased than the base material due to plastic deformation and recrystallization of material at nugget zone. The grain size of SiC and Al_2O_3 reinforced FSW joints is decreased than the unreinforced FSW joints. This is due to fact that the existence of second-phase particles causes the formation of dislocation loops and dislocation cell structure around the particles during deformation, and consequently the grain size decreases as shown in Table 7.10. The grain size is decreased, when the volume proportions of SiC reinforcement increase from 10 to 25% at weld interface, and heat input is increased at weld interface. This is due to developing high temperature at weld interface by increasing volume proportion of reinforcement.

According to Hall–Petch and Orowan Ashby theory, strength of FSW welded joint is dependent on the grain size and interparticle spacing, respectively. So, grain size and interparticle spacing are main factors in controlling the strength of reinforced FS-welded joints. When grain size and interparticle spacing are less, the strengthening is high. Ductility property increases as the grain size and interparticle spacing decrease. Hence, mechanical properties of reinforced friction stir welded joints were enhanced over unreinforced friction stir weld joints [9–11].

The grain size of 30% SiC reinforced friction stir weld of AA6061 is less than the grain size of unreinforced, 10% SiC reinforced and 15% SiC reinforced friction stir welds, but more than the grain size of 25% SiC reinforcement. So mechanical properties were enhanced more at 30% SiC reinforcement compared to 10% SiC reinforced and 15% SiC reinforced friction stir welds, but decreased compared to 25% SiC reinforced friction stir welds. This is because as the heat generated is beyond certain level, the grain size increases. High heat generated causes increase in grain size which lowers the mechanical properties. Also, low heat generated increases interparticle spacing which is lowering the mechanical properties.

Therefore, neither high heat generated (where size of grain large) nor low heat generated (where interparticle spacing is more) guarantees the better mechanical properties of weld. So, select the percentage of volume proportions of reinforcement in such a way that grain size and interparticle spacing are low [9–11].

The grain size of 30% Al_2O_3 reinforced friction stir welds of AA6061 is less than the grain size of unreinforced, 10% Al_2O_3 reinforced, 15% Al_2O_3 reinforced and 25% Al_2O_3 reinforced friction stir welds. This is because of the temperature developed at weld zone of 30% Al_2O_3 reinforcement is more than the temperature developed at weld zone of 10, 15 and 25% reinforcement.

7.4 Conclusions

The following are broad conclusions are drawn from the results.

1. AA 6061 reinforced with silicon carbide and aluminium oxide at different volume fractions of 10, 15, 25 and 30% and without reinforcement were successfully friction stir welded.
2. Non-destructive testing of X-ray radiography and liquid penetrate testing were conducted to check the quality of welded specimens. The test results revealed that all the welds were free from defects.
3. The tensile strength and hardness of FSW portion of AA 6061 are increased with an increase in SiC and Al$_2$O$_3$ reinforcement volume proportions, which had a superior improvement over the unreinforced welds due to the high strength and stiffness of the SiC and Al$_2$O$_3$ reinforcing materials.
4. The tensile strength of Al alloy 6061 with increase in reinforcement volume fraction has resulted in a trend of decreasing ductility. Hence, the unreinforced welded joints have more percentage of elongation against reinforced welded joints.
5. As the reinforcements were added, the specimens failed in bending test because of brittle nature of reinforced particulates.
6. Mechanical properties of SiC reinforced FSW joints of AA 6061 were more enhanced than Al$_2$O$_3$ reinforced FSW joints of aluminium alloy 6061. It was observed that silicon carbide reinforcement is much efficient than aluminium oxide reinforcement.
7. Microstructure evolution of all unreinforced and reinforced friction stir welds is revealed that the grain size of reinforced welded joints at stir zone became fine than unreinforced welded joints grain size due to pinning action which combated the grain growth, and hence the mechanical properties were enhanced.
8. The microstructure of heat-affected zones of all welded specimens has been investigated. The results revealed that the HAZ contained approximately same grain size in all welds. This is because of the fact that the HAZ is only exposed to the welding heat which is not sufficient for deformation and recrystallization of grains.
9. Finally, it is concluded that the FSW process with SiC and Al$_2$O$_3$ particles addition in the weld portion is recommended to enhance the mechanical properties of AA 6061.

References

1. Pradyumn Kumar, A.: A review on friction stir welding for aluminium alloy composite. Int. J. Res. Appl. Sci. Eng. Technol. **3**(11), 324–332 (2015)
2. Indira Rani, M., Marpu, R.N., Kumar, A.C.S.: A study of process parameters of friction stir welded AA 6061 aluminium alloy in O and T6 conditions. ARPN J. Eng. Appl. Sci. **6**(2), 61–66 (2011)

3. Elangovan, K., Balasubramanian, V.: Influences of tool pin profile and tool shoulder diameter on the formation of friction stir processing zone in AA6061 aluminium alloy. Mater. Des. **29**, 362–373 (2008)
4. Karthikeyan, P., Mahadevan, K.: Investigation on the effects of SiC particle addition in the weld zone during friction stir welding of Al 6351 alloy. Int. J. Adv. Manuf. Technol. **80**(9), 1919–1926 (2015)
5. Sun, Y.F., Fujii, H.: The effect of SiC particles on the microstructure and mechanical properties of friction stir welded pure copper joints. Mater. Sci. Eng., A **528**, 5470–5475 (2011)
6. Ceschini, L., Boromei, I., Minak, G., Morri, A., Tarterini, F.: Microstructure, tensile and fatigue properties of AA6061/20 vol.% Al_2O_{3p} friction stir welded joints. Appl. Sci. Manuf. **38**(4), 1200–1210 (2007)
7. Marzoli, L., Strombeck, A.V., Jorge Dos Santos, F., Gambaro, C., Volpone, L.M.: Friction stir welding of an AA6061/Al_2O_3/20p reinforced alloy. Int. J. Compos. Sci. Technol. **66**(2), 363–371 (2006)
8. Aleem Pasha, Md., Ravinder Reddy, P., Laxminarayana, P., Khan, I.A.: The effects of SiC particle addition as reinforcement in the weld zone during friction stir welding of magnesium alloy AZ31B. IRA Int. J. Technol. Eng. **3**(3) (2016). (ISSN 2455-4480)
9. Abbasi, M., Abdollahzaden, A., Bagheri, B., Omidvar, H.: The effect of SiC particle addition during FSW on microstructure and mechanical properties of AZ31 magnesium alloy. J. Mater. Eng. Perform. **24**(12), 5037–5045 (2015)
10. Marzol.: Material flow and thermo-mechanical conditions during friction stir welding of polymers—literature review on experimental results and empirical analysis. Mater. Des. **59**, 344–351 (2014)
11. Kalaiselvan, K., Dinaharan, I., Murugan, N.: Characterization of friction stir welded boron carbide particulate reinforced AA6061 aluminum alloy stir cast composite. Mater. Des. **55**, 176–182 (2014)

Chapter 8
Ultrasonic Spot Welding of Dissimilar Metals: Mechanical Behavior and Microstructural Analysis

Mantra Prasad Satpathy and Susanta Kumar Sahoo

Abstract Ultrasonic welding is considered to be one of the novel and innovative techniques used in semiconductor industries for several years. Recently, it has been a challenging task for the automobile industries to join soft and high thermal conductivity materials like aluminum to copper for lithium-ion battery assembly using conventional fusion welding process. The purpose of this study is to explore the effects of various process parameters like weld pressure, weld time, and vibration amplitude on the weld strength with microstructural analysis of Al–Cu ultrasonic welded joint. The results revealed that the tensile shear and T-peel strength increased with the increase in weld time and attained its maximum at 0.75 s. Afterward, these strengths were gradually decreased. Meantime, different types of microstructures with various properties and morphologies were also observed in the interface zone. The interfacial reaction between Al and Cu produces intermetallic compounds (IMCs) along with swirls and voids. From the energy dispersive X-ray spectroscopy (EDX) and X-ray diffraction (XRD) scans, it was noticed that at high weld time (0.9 s), a 1.5 µm thick IMC layer composed of Al_2Cu and Al_4Cu_9 was formed. A three-dimensional finite element (FE) model was used to find out the compressive stress distribution beneath the sonotrode knurl and the contact stress on the bottom sheet during the delay time. The present work not only illustrates a better understanding of the welding mechanism and failure behavior but also it provides an insight of ultrasonic welding toward the improvement in the quality of weld.

Keywords Ultrasonic welding · Tensile shear strength · T-peel strength Plastic deformation · Intermetallic compound · Microstructure

M. P. Satpathy (✉)
School of Mechanical Engineering, KIIT University,
Bhubaneswar 751024, Odisha, India
e-mail: mantraofficial@gmail.com

S. K. Sahoo
Department of Mechanical Engineering, National Institute of Technology
Rourkela, Rourkela 769008, Odisha, India
e-mail: sks@nitrkl.ac.in

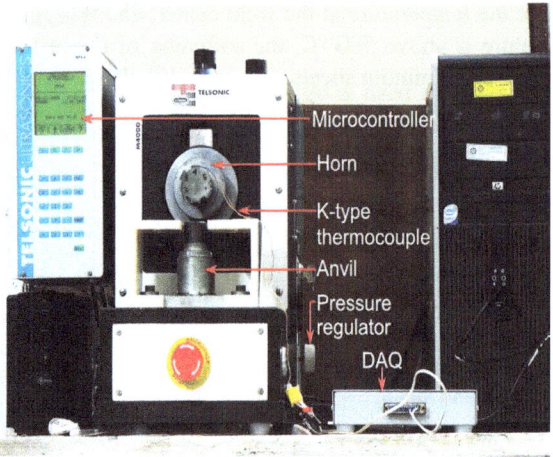

Fig. 8.1 Lateral drive USMW machine setup

Table 8.1 Domain of experiments for USMW of 0.7Al–0.4Cu thickness weld coupons

	Amplitude (μm)	Weld pressure (MPa)	Weld time (s)
Level 1	47	0.34	0.5
Level 2	54	0.38	0.6
Level 3	60	0.42	0.7
Level 4	68	…	0.75
Level 5	…	…	0.8
Level 6	…	…	0.85
Level 7	…	…	0.9

The weld samples of 20×80 mm have been cut from a long sheet in the rolling direction. The welded samples are shown in Fig. 8.2. Meantime, the faying surfaces of the weld specimens have been cleaned with acetone for removing the contaminants and asperities present on it. The physical and mechanical properties of these metals are presented in Table 8.2. The optimum weld strength of the joint has been found out by performing tensile shear and T-peel tests using a universal testing machine INSTRON® 1195. The instantaneous interface temperature is also measured by K-type thermocouples with data acquisition system (DAQ) during the experiments. The exact position of the thermocouple on the weld sample is shown in Fig. 8.3. After welding, transverse sections of the weld coupons have been cut from the parent material and polished with different grades of emery paper and sylvet cloth followed by an etching process for the metallurgical analysis.

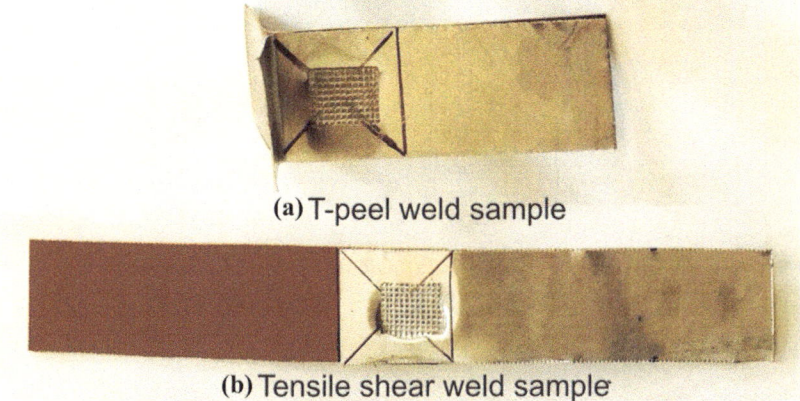

Fig. 8.2 Welded samples for tensile shear and T-peel tests

Table 8.2 Physical and mechanical properties of 0.7Al–0.4Cu weld materials

Properties	Unit	AA1100	UNS C10100
Density	kg/m^3	2710	8940
Young's modulus	GPa	68.9	115
Poisson's ratio		0.33	0.31
UTS	MPa	135.5	302.1
YS	MPa	115.2	251.5

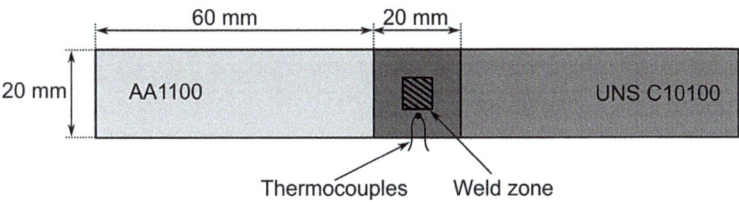

Fig. 8.3 Position of thermocouple for measurement of temperature

8.3 Results of Ultrasonic Spot Welded 0.7Al–0.4Cu Weld Specimen

8.3.1 Mechanical Failure Loads Analysis

After successful completion of welding between Al and Cu specimens by USMW, the weld coupons are detached from each other by a UTM for getting the tensile shear and T-peel ultimate failure loads. These two performance measures with respect to various input parameters like weld pressure and weld time at various amplitudes are illustrated in Fig. 8.4. In this figure, the solid lines represent tensile

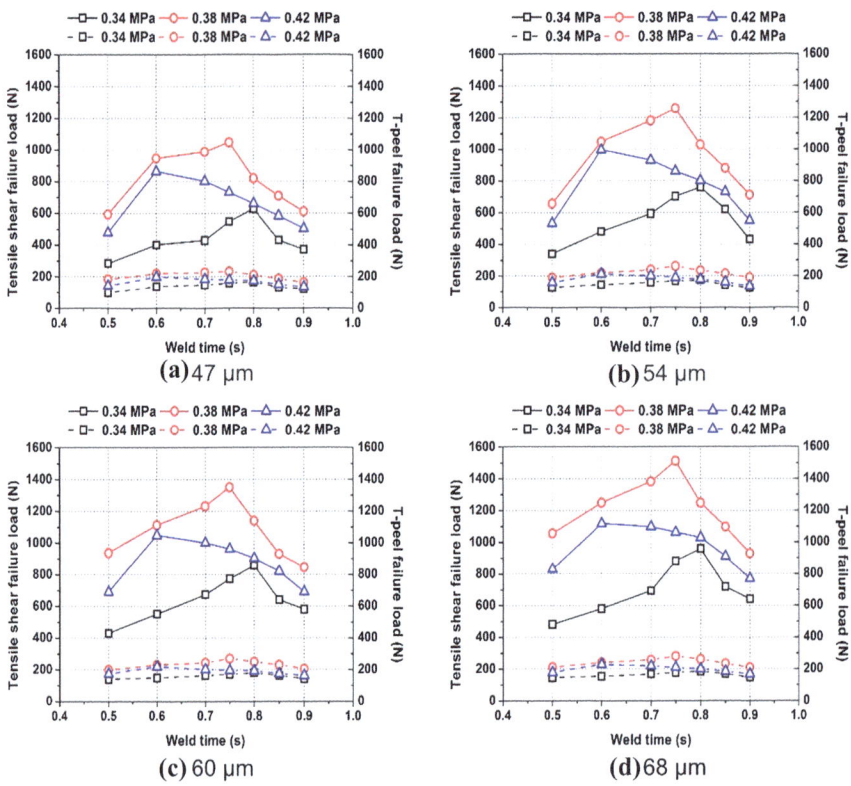

Fig. 8.4 Tensile shear and T-peel failure loads of 0.7Al–0.4Cu weld coupons for different vibration amplitudes

shear failure loads and dotted lines show T-peel failure loads. Meantime, it is noticed that the failure loads increase initially up to a particular value of weld pressure and weld time and afterward, it gradually decreases with the increase of these parameters. Thus, it is expected that the cracks around the weld spot maybe developed due to long weld time. Likewise, when the weld pressure is above than an absolute value, then the relative motion between the sheets hinders and consequently tip sticking and extrusion occur. However, with the increase in vibration amplitude, the relative movement between the sheets increases, and it enhances the plastic deformation rate of sheets. This figure clearly signifies that the highest tensile shear and T-peel failure loads of 1512 and 280.83 N are obtained at 0.38 MPa of weld pressure, 0.75 s of weld time and 68 μm of vibration amplitude. These values are decreased to 1351, 1258, and 1048 N for tensile shear failure loads along with 270.53, 262.17, and 232.92 N for T-peel failure loads at 60, 54, and 47 μm, respectively. It is also observed that for the lower clamping pressure of 0.34 MPa, it takes a longer time to reach the maximum value because the oxide layer may not be broken in such a short period of weld time, and thus, the formation of micro-bonds may not happen. At higher weld pressure of 0.42 MPa, interfacial locking occurs, and due to it, the generated heat breaks the bonds [9].

8.3.2 Weld Area Analysis

During the welding process, the asperities and oxide layer present on the faying surface are dispersed and pure metal to metal contact happens. Thus, it facilitates the formation of micro-bonds in the weld interface region. Moreover, it is very difficult to measure this area during the welding process. Thus, an average weld area is calculated by multiplying the area of one spot with the number of knurls present on the weld tip surface. The knurls on the sonotrode tip make the top surface of sheets nonuniform, and this surface is entirely composed of peaks and valleys. A SEM photograph of the grooved top surface of Al metal is shown in Fig. 8.5. From this figure, it is depicted that the valley region of Al metal is severely plastically deformed due to the lower hardness of Al than Cu and it is the major reason for the formation of micro-bonds along the weld surface. Meantime, the periphery of the valley region withstands severe stress, and thus, the micro-cracks are noticed.

Figure 8.6 illustrates the graph between the average weld areas with various input parameters for 0.7Al–0.4Cu weld coupons. The weld area begins to increase and attains the highest value of 84.52 mm^2 at maximum weld time of 0.9 s, weld pressure of 0.38 MPa, and 68 μm of vibration amplitude. Likewise, comparatively

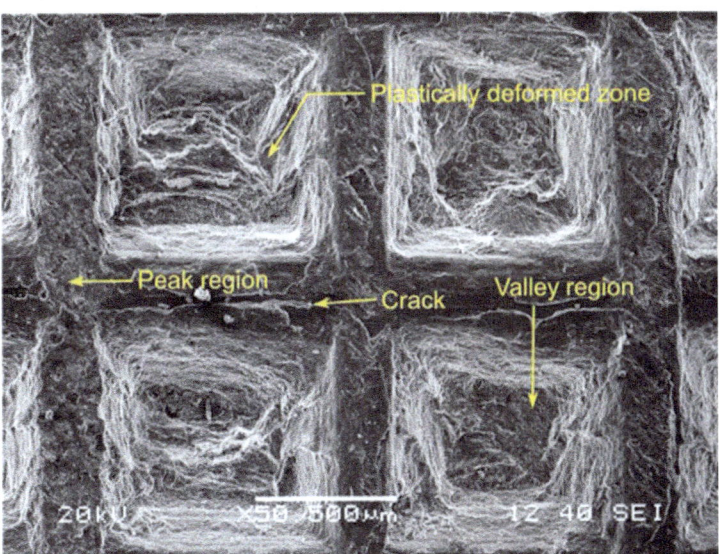

Fig. 8.5 SEM image of top surface of Al sheet under horn tip

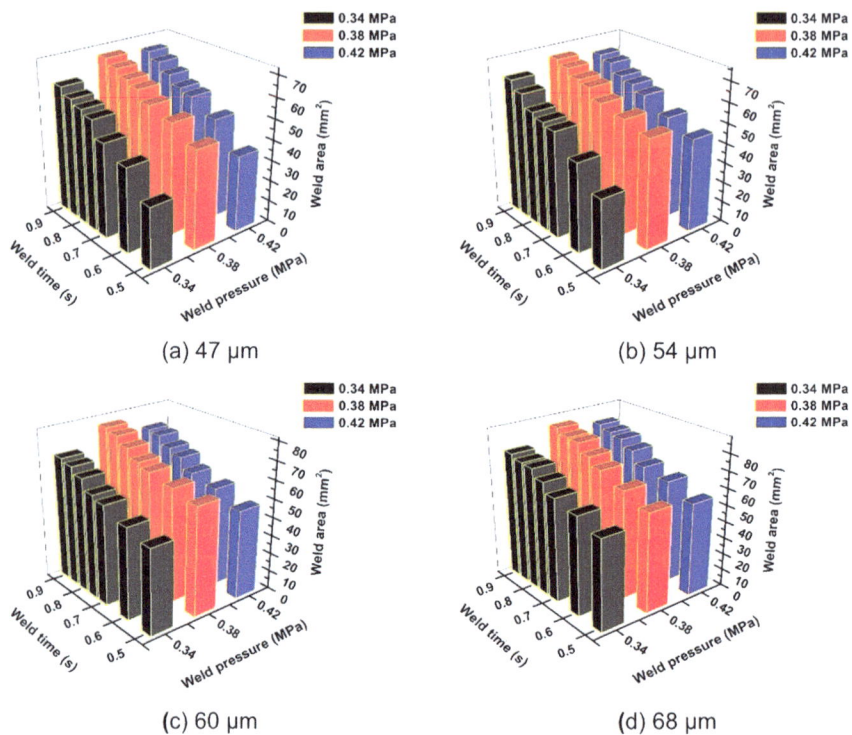

Fig. 8.6 Weld areas of 0.7Al–0.4Cu weld coupons for different amplitudes of vibration

fewer weld areas of 77.63, 71.42, and 69.23 mm^2 are observed for the vibration amplitudes of 60, 54, and 47 μm.

8.3.3 Interface Temperature Analysis

Furthermore, the interface temperature data have been gathered for 0.7Al–0.4Cu samples during the welding, and it is depicted in Fig. 8.7. It can be observed from the figure that there is a sharp increase in temperatures up to 0.6 s for each test condition, and the maximum value of 350 °C is attained in the case of 68 μm of vibration amplitude, 0.9 s of weld time, and 0.38 MPa of weld pressure. However, this interface temperature value decreases to 342.19, 329, and 317 °C for 60, 54, and 47 μm, respectively. As in USMW, plastic deformation plays a predominant role in producing high weld strength and thus, it depends on the amount of scrubbing motion of the sheets. Meantime, it is also noticed that at higher weld

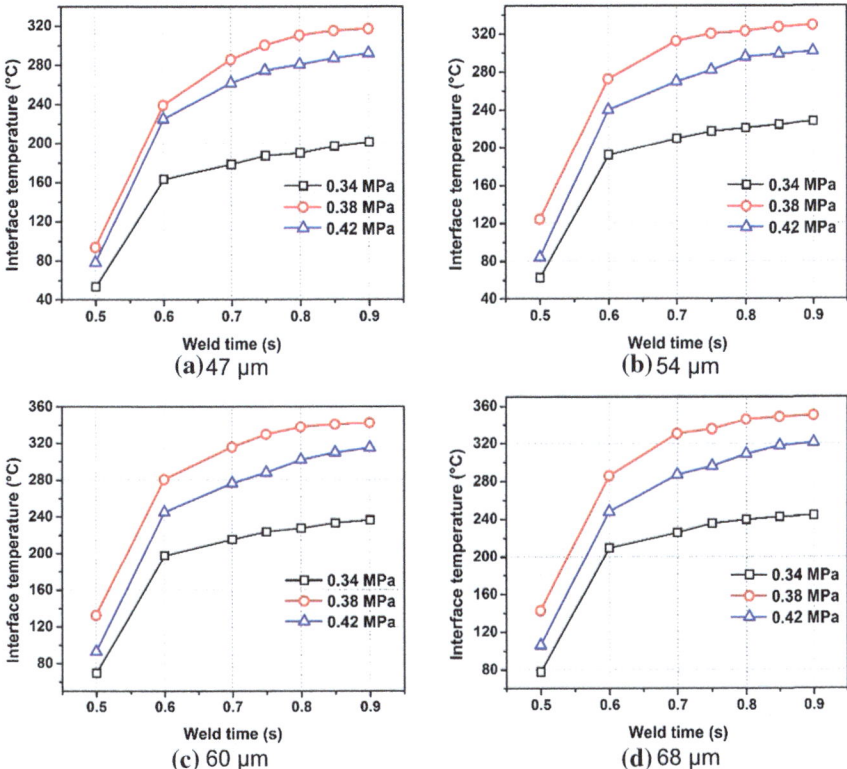

Fig. 8.7 Weld interface temperatures for 0.7Al–0.4Cu weld coupons at different amplitudes of vibration

temperature, the extrusion and drastic reduction of the yield strength of the material happened. Elangovan et al. [10] also experimented with the similar aluminum with copper sheets and predicted the temperatures of nearly 340 °C at the weld zone.

8.4 Finite Element Analysis

Finite element analysis (FEA) is one such promising numerical method that can be used to solve complex problems. In the ultrasonic field, FEA has been applied to determine the vibrational characteristics of the modeled tools prior to its application in the manufacturing industries. It uses a mesh of elements to connect the nodes and also used for modeling of the process. It requires the material properties along with the loading conditions for the simulation purpose and the results that have been found out can be compared with the analytical results.

8.4.1 Compressive Stress Analysis Under Sonotrode Knurls During Delay Time

At the starting time of the weld, the sonotrode knurls pressed the specimens firmly without any vibration until the delay time is over. For the analysis point of view, 0.2 MPa clamping pressure is used. This penetration is necessary to provide adequate friction during the welding process. Thus, to find out the compressive stress exerted by the sonotrode on the top sheet, FEA program ANSYS® is used. Meantime, the maximum depth of indentation depends on the height of the sonotrode knurls, and it is taken as 0.2 mm. Figure 8.8 depicts the schematic model of a total system which is used for further analysis. Due to less welding time, relatively less thermal conductive sonotrode and anvil material, and for simplification of the 3D problem, only half of the total system is taken into consideration.

For doing this analysis, static structural module of ANSYS® is used with practical boundary conditions. As the correctness of the FEA solution depends on the mesh element size and its types and thus, hexahedral meshing is selected for sonotrode and two sheets along with tetrahedral meshing for an anvil. The number of mesh elements varies from one part to another part and the finest mesh size is chosen for the weld interface as guided by previous work done in friction stir welding [11]. The material properties included in the model are Young's modulus, Poisson's ratio, specific heat, thermal conductivity and density [12]. It is already furnished in Table 8.2. Figure 8.9 illustrates the meshing for the whole system. The boundaries of this model are considered as frictionless rigid walls to find out the stress distribution beneath the sonotrode knurls.

The deformation and normal stress distribution on the weld samples along the normal (Y-direction) are shown in Fig. 8.10. It is observed from Fig. 8.10a that due to the penetration of sonotrode knurls, the deformations at the weld spots are obtained with a range from 253×10^{-6} to 327×10^{-6} m. Likewise, Fig. 8.10b reveals that the normal stress distributions are in the range from 0.46 to 0.58 MPa at

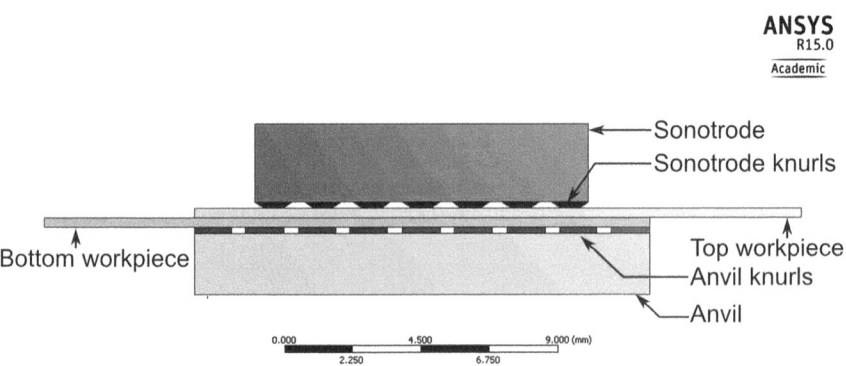

Fig. 8.8 Schematic model used for FE analysis

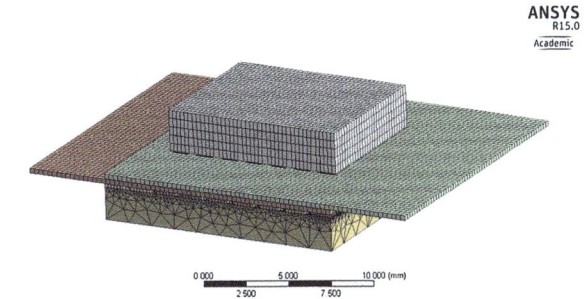

Fig. 8.9 Illustrating various mesh elements used in proposed FE model

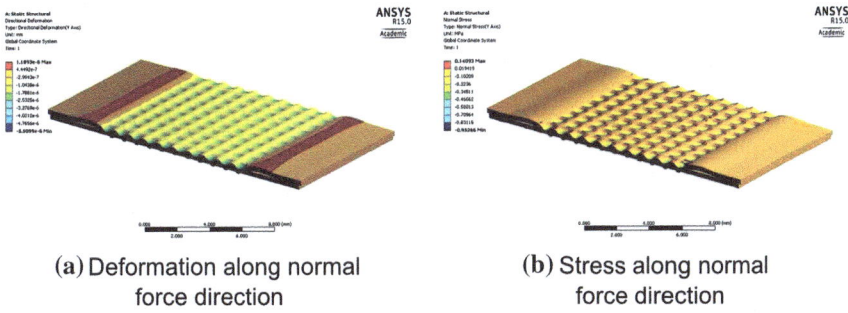

(a) Deformation along normal force direction

(b) Stress along normal force direction

Fig. 8.10 Deformation and stress distributions of 0.7Al–0.4Cu weld coupons along normal direction (*Y*-axis) during delay time

the centers of the weld spots. It is also observed from the figure that the noncontact portions withstand minimum compressive stress.

The compressive stress for one weld spot is represented in Fig. 8.11 with lines which are parallel to the normal force direction. It indicates up to 220 μm beneath

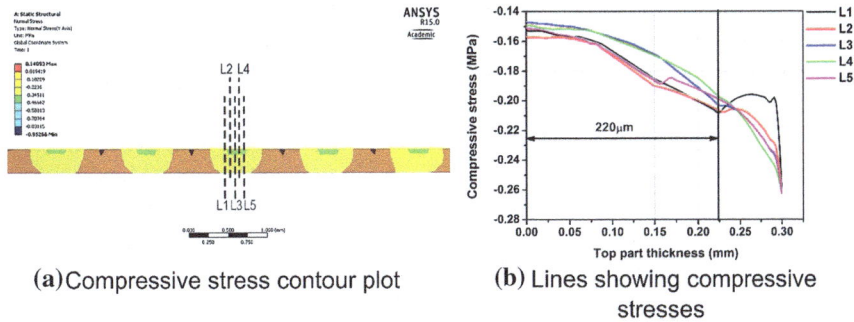

(a) Compressive stress contour plot

(b) Lines showing compressive stresses

Fig. 8.11 Compressive stress distributions along lines parallel to normal force direction for one weld spot

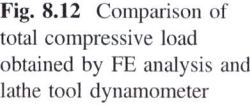
Fig. 8.12 Comparison of total compressive load obtained by FE analysis and lathe tool dynamometer

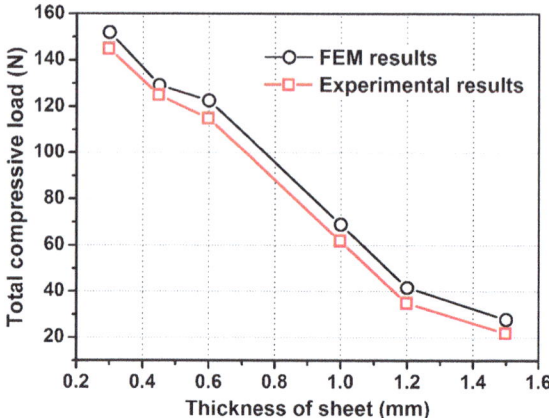

the sonotrode knurls; the compressive stress is decreasing in a uniform manner. But after that, the changes are erratic in nature. It means up to 220 μm thickness of the top part, the compressive force has some effect.

Figure 8.12 presents the comparison graph of total compressive load obtained by FE analysis and lathe tool dynamometer. It can be clearly observed that the FE analysis results show a similar nature of curves as of the experiments. Thus, it can be said that the present finite element model is capable of predicting the compressive stress distribution appreciably.

8.4.2 Effect of Top Part Thickness on Contact Stresses at Weld Spot

In this section, the effect of top part thickness on the contact stresses between the sheets is explored through FE analysis. The importance of this study is to determine the maximum thickness of the sheets that can be weldable using the current ultrasonic welding machine. Thus, a nominal clamping pressure of 0.2 MPa is applied during the analysis. In this case, the rectangular knurled sonotrode tip is pressed to the Al–Cu sheets, which in turn fixed on a circular flat anvil. The area of the anvil is greater than the area of weld tip. The peak compressive stress distributions on the bottom sheet (Cu) are illustrated in a graphical manner (Fig. 8.13).

It can be clearly observed from the figure that the maximum contact stress is significantly decreased on the increase of sheet thickness. It is also revealed that for a top sheet (Al) thickness up to 0.6 mm, the bottom worksheet is experienced almost uniform compressive stress in the weld zone. But when the thickness is increased further, and above 1 mm thickness of the workpiece, the peak contact stress is drastically reduced. In real welding condition also, it is not possible to

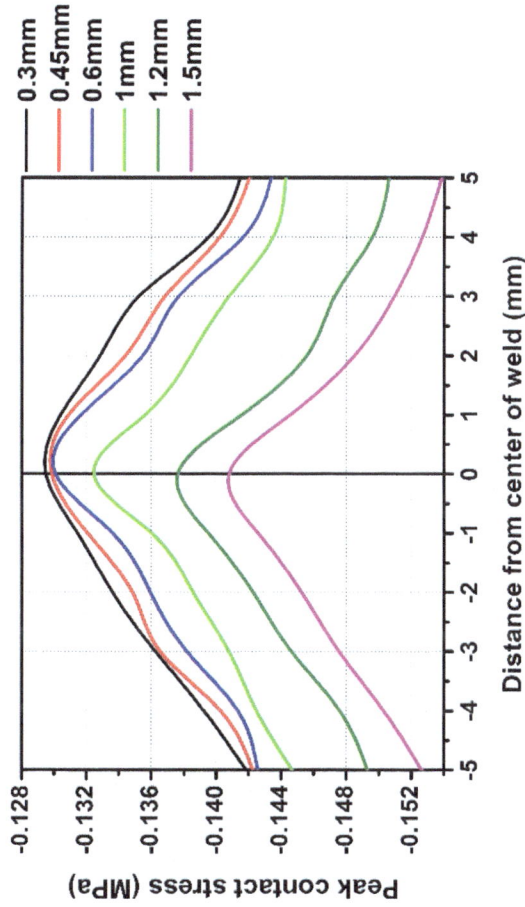

Fig. 8.13 Distribution of peak contact stress on bottom sheet for various thickness of top part

achieve the welds beyond 1-mm-thick sheet at these machine settings. These results show somewhat similar nature as the results given by Rozenberg et al. [13].

8.5 Metallurgical Analysis

The weld quality not only depends on the proper selection process parameters but also depends on the material properties and thickness of the workpiece. At higher weld time and vibration amplitude, the interatomic diffusion rate is increased, and plastic flow occurs at the weld interface [14]. SEM images along the cross sections of 'under,' 'good' and 'over' weld conditions using 0.7Al–0.4Cu weld coupons are shown in Fig. 8.14. In under weld condition, there is a gap between two materials

Fig. 8.14 SEM images of weld cross sections for 0.7Al–0.4Cu weld coupons

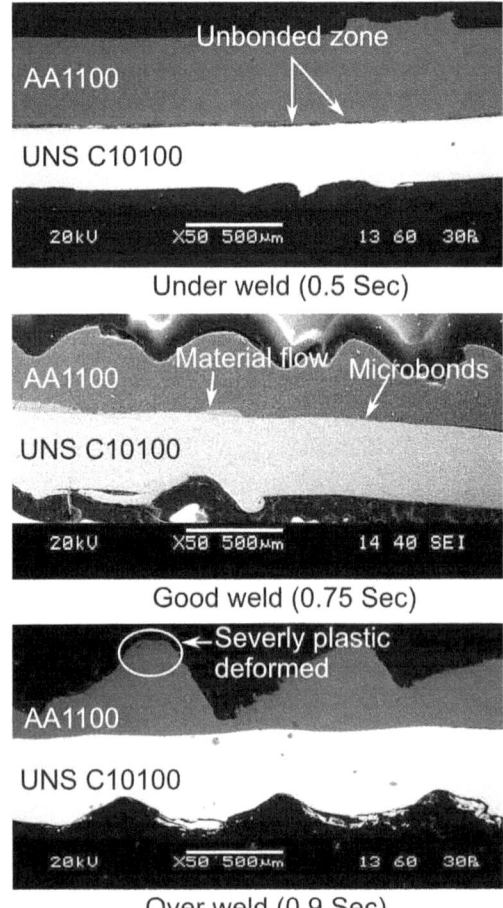

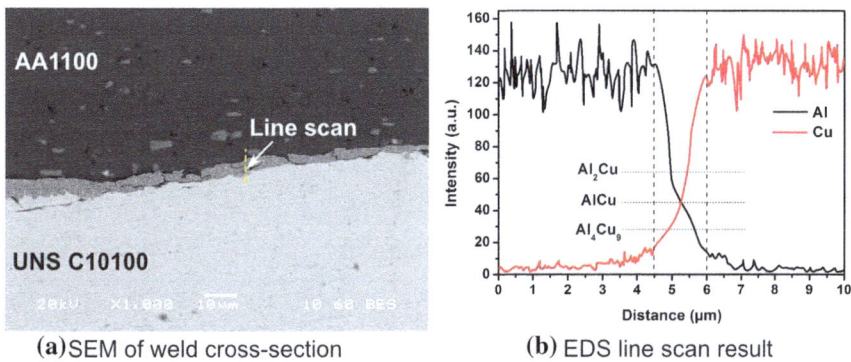

(a) SEM of weld cross-section   (b) EDS line scan result

Fig. 8.15 **a** High-magnification SEM image and **b** EDS line analysis of weld interface for good weld joint

observed with no micro-bonds present in it while, in good condition, the welding interface is highly packed with micro-bonds. The bond density is comparatively high in good weld condition. But in over weld condition, there is no continuous weld interface observed due to severe plastic deformation of the aluminum material. In this condition, as the temperature increases, the sheets get to be more ductile and subsequently profound extrusion happens with tip sticking of the top sheet.

The SEM micrograph with EDS mapping is shown in Fig. 8.15. The EDS line scan results reveal that there is no continuous IMC formed in between the two sheets due to quick process and comparatively low-temperature generation. The diffusion interlayer thickness is observed as about 1.5 μm, and the chemical compositions (wt%) is 66.67Al–33.33Cu. According to the phase diagram of Al–Cu, the possible IMC formed maybe composed of AlCu, Al_2Cu, and Al_4Cu_9. Moreover, as the interface temperature surpasses 350 °C and thus, it suggests that the dominant IMC is probably Al_2Cu according to the Al–Cu phase diagram [15].

Meantime, the micro-hardness at the weld interface is HV88.55, and it is more than the micro-hardness of the Al base material (HV59.55). Furthermore, XRD analysis also has been performed to validate the results obtained from EDS. Thus, XRD analysis has been done on both the fractured surface of Al and Cu. Figure 8.16 depicts the peak intensities of various compounds with respect to diffraction angle 2θ. From this graph, it is noticed that only one IMC is formed, and that is Al_2Cu. Thus, it is the primary reason for lowering the tensile shear and T-peel failure loads. A similar type of results is also reported by other researchers [16, 17] in USMW.

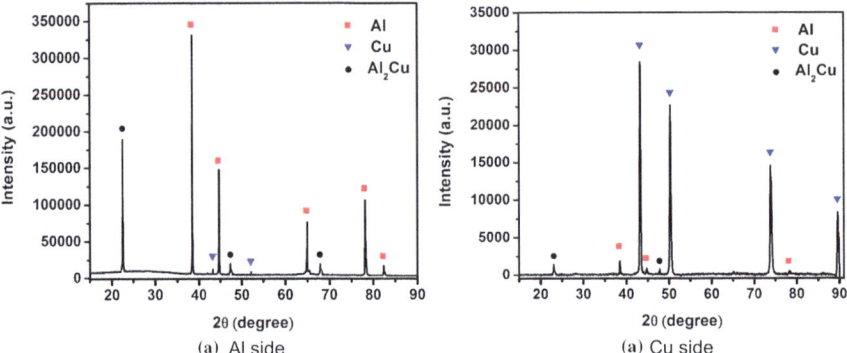

Fig. 8.16 XRD results showing the presence various elements on **a** Al fractured surface, **b** Cu fractured surface

8.6 Conclusions

The present study covers all the aspects of USMW for Al–Cu dissimilar materials. This study not only gives the information about joint strengths, interface temperatures, and weld areas but also reveals the quality of weld through metallurgical investigation. The following outcomes can be deduced from this research:

- Al and Cu sheets can be welded successfully under different process parameter combinations. From the experimental research, it was observed that during the welding of these dissimilar sheets, the highest tensile shear and T-peel failure loads of 1512 and 280.83 N were obtained for 0.7Al–0.4Cu at 0.38 MPa weld pressure, 0.75 s of weld time, and 68 μm of vibration amplitude.
- The weld area analysis revealed that for the 68 μm of vibration amplitude, a weld time of 0.9 s, and weld pressure of 0.38 MPa, the highest weld area of 84.52 mm^2 is obtained.
- Likewise, the interface temperature increased with the weld time and attained the highest temperature of 350 °C at 68 μm of vibration amplitude, 0.9 s of weld time, and 0.38 MPa of weld pressure values.
- The microscopic analysis also has been done for 0.7Al–0.4Cu because it provides the solid information about the maximum tensile shear and T-peel failure loads for the weld joint. Thus, three quality characteristics, viz., 'under weld,' 'good weld' and 'over weld' are proposed with respect to the formation of micro-bonds.
- The numerical model can predict the compressive stress under the knurl pattern and the thickness of the sheets. The compressive stress is decreasing in a uniform manner up to 220 μm thickness of the top part, and the FE analysis results show a similar trend as of the experiments. Furthermore, when the top part thickness is above 1 mm thickness, the peak contact stress is drastically reduced, and it is not possible to weld these materials.

- Again, the presence of IMCs is confirmed on either side of the fracture surface through EDS analysis. The result also revealed that there is a formation of about 1.5-μm-thick Al_2Cu IMC layer for the good weld joint. Meanwhile, XRD analysis is also carried out to validate the EDS results.

References

1. Bakavos, D., Prangnell, P.B.: Mechanisms of joint and microstructure formation in high power ultrasonic spot welding 6111 aluminium automotive sheet. Mater. Sci. Eng., A **527** (23), 6320–6334 (2010)
2. Komiyama, K., Sasaki, T., Watanabe, Y.: Effect of tool edge geometry in ultrasonic welding. J. Mater. Process. Technol. **229**, 714–721 (2016)
3. Sasaki, T., Hosokawa, Y.: Effect of relative motion between weld tool and work piece on microstructure of ultrasonically welded joint. Mater. Sci. Forum, 1782–1787 (2014)
4. Shakil, M., Tariq, N.H., Ahmad, M., Choudhary, M.A., Akhter, J.I., Babu, S.S.: Effect of ultrasonic welding parameters on microstructure and mechanical properties of dissimilar joints. Mater. Des. **55**, 263–273 (2014)
5. Zhang, C.Y., Chen, D.L., Luo, A.A.: Joining 5754 automotive aluminum alloy 2-Mm-thick sheets using ultrasonic spot welding. Weld. J. **93**, 131–138 (2014)
6. Seo, J.S., Jang, H.S., Park, D.S.: Ultrasonic welding of Ni and Cu sheets. Mater. Manuf. Process. **30**(9), 1069–1073 (2015)
7. Norouzi, A., Hamedi, M., Adineh, V.R.: Strength modeling and optimizing ultrasonic welded parts of ABS-PMMA using artificial intelligence methods. Int. J. Adv. Manuf. Technol. **61**(1–4), 135–147 (2012)
8. Siddiq, A. M., Ghassemieh, E.: Modelling and characterization of ultrasonic consolidation process of aluminium alloys. In: Materials Research Society Symposium Proceedings, pp. 125–132 (2008)
9. Harthoorn, J.L.: Ultrasonic Metal Welding. Technische Hogeschool Eindhoven (1978)
10. Elangovan, S., Semeer, S., Prakasan, K.: Temperature and stress distribution in ultrasonic metal welding—an FEA-based study. J. Mater. Process. Technol. **209**(3), 1143–1150 (2009)
11. Jedrasiak, P., Shercliff, H.R., Chen, Y.C., Wang, L., Prangnell, P., Robson, J.: Modeling of the thermal field in dissimilar alloy ultrasonic welding. J. Mater. Eng. Perform. **24**(2), 799–807 (2015)
12. Khanna, S.K., Long, X., Porter, W.D., Wang, H., Liu, C.K., Radovic, M., Lara-Curzio, E.: Residual stresses in spot welded new generation aluminium alloys Part A—thermophysical and thermomechanical properties of 6111 and 5754 aluminium alloys. Sci. Technol. Weld. Join. **10**(1), 82–87 (2005)
13. Rozenberg, L.: Physical Principles of Ultrasonic Technology. Springer Science & Business Media (2013)
14. Al-Sarraf, Z., Lucas, M.: A study of weld quality in ultrasonic spot welding of similar and dissimilar metals. J. Phys. Conf. Ser., 12013 (2012)
15. Funamizu, Y., Watanabe, K.: Interdiffusion in the Al–Cu system. Trans. Japan Inst. Met. **12**(3), 147–152 (1971)
16. Balasundaram, R., Patel, V.K., Bhole, S.D., Chen, D.L.: Effect of zinc interlayer on ultrasonic spot welded aluminum-to-copper joints. Mater. Sci. Eng., A **607**, 277–286 (2014)
17. Yang, J.W., Cao, B., He, X.C., Luo, H.S.: Microstructure evolution and mechanical properties of Cu-al joints by ultrasonic welding. Sci. Technol. Weld. Join. **19**(6), 500–504 (2014)

Chapter 9
Distribution of Electromagnetic Field and Pressure of Single-Turn Circular Coil for Magnetic Pulse Welding Using FEM

Mohammed Rajik Khan, Alok Raj, Md. Mosarraf Hossain, Satendra Kumar and Archana Sharma

Abstract Magnetic pulse welding (MPW), which is uniquely advantageous in welding electrically conductive similar and dissimilar pipe fittings, is a contactless welding technology based on high-speed magnetic impulse shaping and solid-phase diffusion welding. This has proven to be an effective solution to specific manufacturing problems, especially for leak-proof dissimilar pipe joints required to sustain high pressure, which is very difficult to achieve by conventional techniques. For achieving the successful weldament, it is essential to understand the effect of various process parameters to generate proper weldability window. In the present work, the distribution of electromagnetic force and magnetic field of single-turn circular coil for MPW has been investigated using FE simulation. A three-dimensional (3D) electromagnetic FE model has been developed using commercially available ANSYS-EMAG application software. A single-turn inductor coil of Cu material is chosen for the analysis. Compression joining of tubular metallic assembly with flyer tube as Al6061 and target tube as SS304 is simulated in ANSYS Maxwell 3D to study the influence of varying process

M. R. Khan (✉) · A. Raj · Md. M. Hossain
Department of Industrial Design, National Institute of Technology Rourkela,
Rourkela 769008, India
e-mail: khanmr@nitrkl.ac.in

A. Raj
e-mail: alokraj169@gmail.com

Md. M. Hossain
e-mail: mosarraf.world@gmail.com

S. Kumar · A. Sharma
Accelerator and Pulse Power Division, Bhabha Atomic Research Center,
Mumbai 400085, India
e-mail: satendra@barc.gov.in

A. Sharma
e-mail: arsharma@barc.gov.in

© Springer Nature Singapore Pte Ltd. 2019
U. S. Dixit and R. G. Narayanan (eds.), *Strengthening and Joining by Plastic Deformation*, Lecture Notes on Multidisciplinary Industrial Engineering,
https://doi.org/10.1007/978-981-13-0378-4_9

parameters like air gap between the tubes, tube thickness, and gap between the flyer tube and the coil with respect to the input voltage. FE simulation results for weld formation are verified with the analytical results and the data available in the literature. The study reveals that for effective welding, estimation of electromagnetic field and electromagnetic force has a significant role which is governed by the process parameters of applied voltage and air gap. The presented information can assist to understand process physics, coil reliability, and prediction of mechanical behavior of the workpiece.

Keywords Magnetic pulse welding · FE model · MPW coil · Magnetic field and pressure distribution

9.1 Introduction

Magnetic pulse welding (MPW)/electromagnetic welding (EMW) is a unique and reliable technique for joining similar and dissimilar workpiece arrangement fulfilling the present industrial expectation of maintaining high strength to weight ratio. Various welding processes presently practiced in the industry are not capable of providing perfectly leak-proof welded joint and require various additional reworks to get the final job [1, 2]. Light weighted parts obtained from the coalescence of different materials are highly postulated in aerospace, automobile, and nuclear power industries [3]. The dissimilar material joint can only be conceived by pulse welding techniques with smooth and cleaner joint interface. It is a heat-free solid-state joining process which eliminates localized annealing and good for workpiece with length scale in the order of millimeter to centimeter. Despite negligible process failure and high production rate, MPW has limited industrial use due to lack of guidelines regarding process parameters affecting the whole welding process.

   Over the past decade, many researchers have shown keen interest toward coalescence of dissimilar metals. Guo et al. [4] have performed explosive welding to evaluate the plastic deformation of aluminum and SS bimetal clad pipes. Tabbataee and Mahmoudi [5] have performed FE simulation of explosion welding and conducted experimentations for the joining of titanium and stainless steel cladding plates. Mousavi and Joodaki [6] have performed experimentations of explosive welding to join Al5056, Al1015, and SS304 tubes in a single processing step. Numerical simulations are developed to predict the distribution of magnetic field and force in electromagnetic-forming systems [7, 8]. An experimental investigation on the joining of multiple sheet metal combinations is known through the work done in the past [9–12]. A practical weldability window to weld aluminum alloy (Al6060T6) tubular assembly by MPW has been investigated by Raoelison et al. [13]. The weld quality assessment was based on the fracture surface obtained by a destructive bonding test and macrographic analysis of welded joint section. A three-dimensional (3D) electromagnetic FE model has been developed to analyze the electromagnetic force distribution using square-type working coil by Shim and Kang [14]. Results of numerical analysis were compared with experimental results, and leakage test was conducted to verify the weldability of Al1070, SM45C for Al and steel square pipe and rod, respectively. Xu et al. [15] studied the effect of impact velocity in magnetic impulse welding for Al–Fe pipe fittings under varying voltages and presented the combined results of numerical simulation and technological test. The joint strength and wave interface were also analyzed by peeling test and microstructure analysis. Cui et al. [16] investigated experimentally the critical thickness of the inner tube required for Al–Fe tubular joints in MPW. Kapil and Sharma [17] have presented sequential coupling of electromagnetic–structural simulation for MPW of tubular jobs. Based on the change of input voltage and air gap between structural steel ASTM A36 tubular workpieces, an optimal weldability window has been suggested. Analytical model to predict generated magnetic pressure and velocity gained by the outer tubular workpiece in a single-turn axisymmetric coil along with proper experimental verification for electromagnetic forming and welding is presented by Nassiri et al. [18].

Till date, very limited research work has been reported for EMW of tubular arrangements, and no major attempts were made in the standardization of process parameters. Here, authors have presented a 3D electromagnetic FE simulation for MPW of tubular joints with varying air gap and input voltage. A single-turn Cu coil with Al6061 as flyer tube and SS304 as the target tube is considered for the analysis. Distribution of electromagnetic force and field is simulated with varying input process parameter to identify essential process parameter values required for successful welding. It is observed that magnetic field, surface force density, and current distribution change by varying the specific process parameters like standoff distance and input voltage, which drastically affects the welding performance.

9.2 Electromagnetic Welding

Electromagnetic Welding is an effective and successful high-speed welding technology [19] that uses electromagnetic force to join two similar or dissimilar metals and nonmetals. The joining mechanism is almost similar to that of explosive welding [4–6]. High-velocity collision and shear strain generated at the bonding interface leads to a strong solid-state weld undergoing minimal intermetallic phase transformation at the collision zone.

9.2.1 Principle

EMW technology is characterized as a very short welding process for joining similar/dissimilar joints, which is accomplished within few microseconds. The process begins at the capacitor bank which is charged to a certain voltage to fulfill the energy required for welding. The workpieces are then placed in the vicinity of the electromagnetic coil and the capacitor is then discharged by the trigger circuit. Consequently, a large current starts flowing through the attached coil. The circulating current in the coil creates a changing magnetic flux, which in turn induces a secondary current on the outer workpiece. A secondary magnetic field of opposite nature is generated due to the secondary current, i.e., eddy current induced on the outer tube. The two magnetic fields in the opposite nature lead to a repulsive magnetic force, called Lorentz force which causes the flyer to collide with the base workpiece under very high velocity. The Lorentz force [10] exerted on the outer workpiece is defined as

$$F = J \times B, \tag{9.1}$$

where J is the current density (in A/m^2) and B is the magnetic flux density (in T). The schematic layout of electromagnetic welding system [20] is shown in Fig. 9.1.

9.2.2 Process Modeling

The process modeling gives the required specific values of process parameters governing the EMW process for possible weld between workpieces. A step-by-step analytical approach is presented to estimate the process parameters in EMW of tubular fittings. The process modeling starts with an initial assumption of flyer tube velocity. The required acceleration [21] of the flyer tube can be expressed as

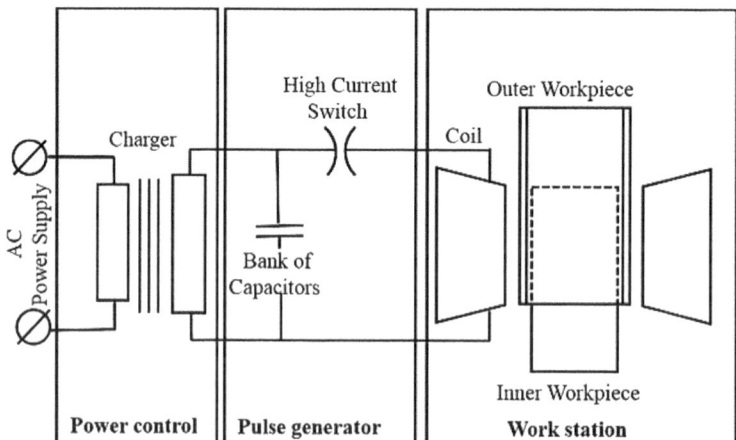

Fig. 9.1 Schematic of electromagnetic welding system

$$a = \frac{V_c^2}{2s}, \tag{9.2}$$

where V_c is the collision velocity of the flyer tube and s is the air gap between the flyer tube and the base tube. Magnetic pressure [21] required for EMW can be obtained as

$$P_{\text{mag}} = P_{\text{def}} + P_{\text{acc}}. \tag{9.3}$$

Again,

$$P_{\text{def}} = \frac{2t\sigma_{\text{uts}}}{R} \tag{9.4}$$

and

$$P_{\text{acc}} = \frac{t\rho V_c^2}{2s}, \tag{9.5}$$

where P_{def} and P_{acc} are the deformation and acceleration pressure, respectively, t and R are the thickness and average radius of the outer tube, respectively, and σ_{uts} and ρ are the ultimate tensile strength and the density of the flyer tube material, respectively. The magnetic field developed in the vicinity of the coil can be obtained [21] as

$$B = \sqrt{2\mu P_{\text{mag}}}, \tag{9.6}$$

where μ is the magnetic permeability of the flyer tube material. Again $\mu = \mu_r \mu_0$, where $\mu_r = 1$ is the relative permeability of the flyer tube material and $\mu_0 = 4\pi \times 10^{-7}$ is the magnetic permeability of space. Using relation $B = \mu_0 H$, the input current can be defined as

$$I = Hl_0/N, \tag{9.7}$$

where H is the magnetic field intensity developed in the coil, and N is the number of turns in the coil and l_0 is the width of the coil. The inductance of the coil can be defined by the ideal solenoid expression [22] as

$$L = \frac{\mu_0 \pi r_{\text{eff}}^2}{l_i}, \tag{9.8}$$

where r_{eff} is the effective radius of the coil and l_i is the landing length of the coil. The input current can be expressed [18] as

$$I(t) = V\sqrt{\frac{C}{L}} \times e^{-\zeta\omega t} \times \frac{\sin(\omega\sqrt{1-\zeta^2}t)}{\sqrt{1-\zeta^2}}, \tag{9.9}$$

where V, ζ, ω and t are the input voltage, damping ratio, natural frequency, and process time, respectively. The skin depth up to which current can penetrate is taken [7] as

$$\delta = \sqrt{\frac{2\rho}{\mu_r\mu_0\omega}}, \tag{9.10}$$

where μ_r and μ_0 are the relative magnetic permeability and permeability of space.

9.2.3 EMW Coil Modeling

The axisymmetric EMW coil generates a huge magnetic field in its vicinity causing the requisite velocity of the flyer tube/plate. Its distribution strongly depends on the coil geometry and the material. Hence, to get uniform and maximum field distribution, a better coil design is essential. To enhance and concentrate the field toward the weld zone, a field concentrator with converging shape is considered. For coil modeling, first of all, the tapered part generally called as the field concentrator is designed with few basic assumptions as

- Landing length of the coil is equal to the required weld width.
- Gap between the flyer tube and the inner radius of the coil (h) is constant.
- Angle of collision of flyer tube with base tube be $0°$.
- Taper angle (θ) of field concentrator be $45°$.

To effectively design a coil, the two most important parameters are the coil effective radius and the coil efficiency. The generic definition to determine the efficiency [21] and effective radius [22] is defined, respectively, as

$$\varepsilon = \frac{l_0}{l_i + \frac{a\sin\theta}{1-\cos\theta}} \tag{9.11}$$

and

$$r_{\text{eff}} = \left[1 - \left(\frac{r_w}{r_a}\right)^2\right], \tag{9.12}$$

where ε is the efficiency of the coil, a is the distance between the axis of the coil and the field concentrator, $r_w = (r_o - r_i)/2$, $r_a = r_i + h + r_w$ and r_o, r_i and r_a are the outer, inner, and average radius of the coil. The pictorial view of field concentrator is depicted in Fig. 9.2. Figure 9.3 shows the orthographic view of the axisymmetric coil.

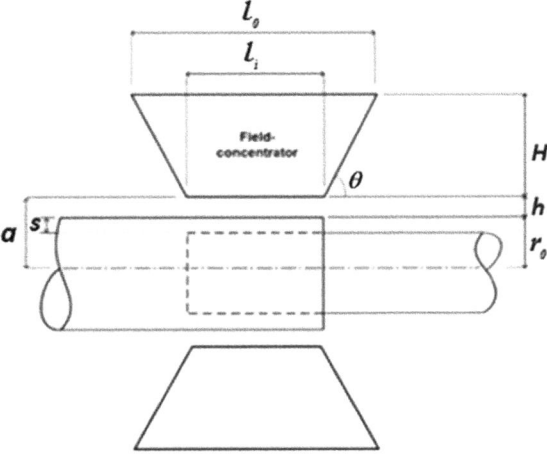

Fig. 9.2 Conical field concentrator with flyer tube

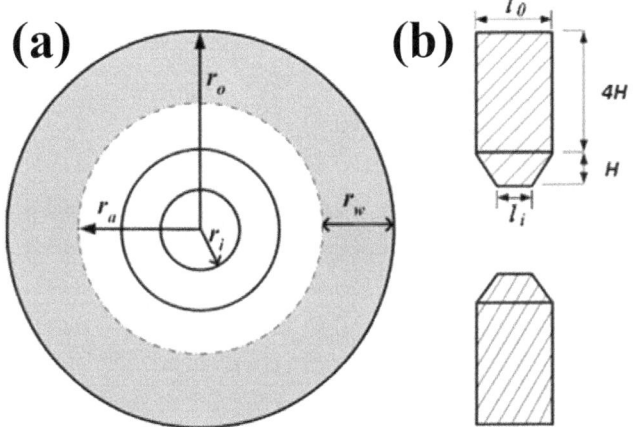

Fig. 9.3 Orthographic view of the EMW coil: **a** front view **b** side cross—sectional view

9.2.4 Design of EMW Coil

The coil design is one of the major parts that influence the work process parameters. The shape of the coil is a factor which decides the pressure distribution on the flyer workpiece. For the EMW process of tubular jobs, single-turn and multi-turn coils are mostly applied. A multi-turn coil can be used with the combination of several field shapers suitable for different geometries. However, a well-designed single-turn coil is efficient as more energy is dissipated in the field shaper.

Table 9.1 Dimensional parameters of EMW coil

Dimensional parameters	Values
Width of the coil (l_0)	32 mm
Landing length of the coil (l_i)	10 mm
Inner radius of the coil (r_i)	10.5 mm
Outer radius of the coil (r_o)	65.5 mm
Taper angle of the field concentrator (θ)	45°
Height of the field concentrator (H)	11 mm
Gap between the flyer tube and the inner radius of the coil (h)	0.5 mm

The coil geometry is dependent on the tube's material to be welded and welding process parameters. Here, the single-turn coil is designed as per the relations discussed in Eqs. (9.11) and (9.12) considering weld width (l_i) of 10 mm, flyer tube collision velocity of 290 m/s, and coil material as copper with an efficiency (ε) of 90%. Other dimensional parameters required to generate the 3D geometry of coil are mentioned in Table 9.1. The 3D CAD model of the proposed coil is shown in Fig. 9.4. Here, SS304 and Al6061 were selected as materials of the inner and outer tube. The welding arrangement typically composed of hollow flyer and base tubes. The outer and inner diameters of the flyer tube were kept fixed at 20 mm and 18 mm, respectively. Table 9.2 lists the geometric dimensions of base tube (SS304) with three different standoff distances considered in FE simulation. Mechanical properties of flyer tube (Al6061) are shown in Table 9.3.

Fig. 9.4 3D CAD model of the proposed working coil

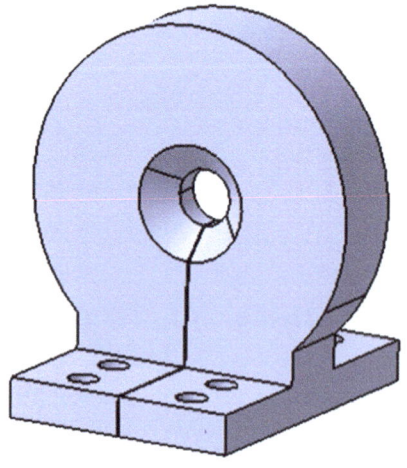

Table 9.2 Geometric dimensions of base tube (SS304) considered for FE simulation

Material	Outer diameter (mm)	Inner diameter (mm)	Air gap (mm)	Thickness (mm)
SS304	17	14	0.5	1.5
	16.5	13.5	0.75	1.5
	16	13	1	1.5

Table 9.3 Mechanical properties of flyer tube Al6061-T6

Property	Value
Density (ρ)	2700 kg/m^3
Ultimate tensile stress (σ_{uts})	310 MPa
Yield stress (σ_{yield})	276 MPa
Modulus of elasticity (E)	68.9 GPa
Speed of sound (c)	3150 m/s
Poisson's ratio (v)	0.33

9.3   FE Simulation for Joining of Al6061 and SS304

The finite element simulation for the EMW of Al6061 and SS304 tubular jobs is carried out using ANSYS Maxwell 3D. Here, air gaps considered are 0.5, 0.75, and 1 mm. The input voltage varies in steps of 1 kV from 15 to 21 kV. Taking the welding parameters from Table 9.4, the simulation is performed to identify the magnetic field and force with varying input parameters (air gap and voltage). The simulation is conducted for 30 μs with time step of 1 μs. For the present run, the peak current reaches 4 μs. Distribution of magnetic field, surface force density, and peak current is recorded for the simulation. Figure 9.5 shows the magnetic field distribution for EMW with an air gap of 1 mm at 4 μs and 16 kV input energy. Distribution of surface force density, magnetic field, and peak current over flyer tube with varying voltages at multiple air gaps are shown in Figs. 9.6, 9.7 and 9.8, respectively.

Table 9.4 Welding setup parameter

Welding parameters	Values
Inductance of the coil (L)	67.725 nH
Effective radius of the coil (r_{eff})	13.0977 mm
Capacitance of the capacitor bank (C)	108 μF
Resistance of the coil (R)	0.2494 mΩ
Damping ratio (ζ)	4.991×10^{-3}
Frequency (ω)	3.7×10^{5} rad/s
Efficiency of the coil (ε)	90%
Width of the coil (l_0)	32 mm
Skin depth of the coil (δ)	0.4 mm

Fig. 9.5 Magnetic field distribution in EMW of Al6061T6 and SS304 tubular jobs at an air gap of 1 mm at 4 μs and 16 kV input energy

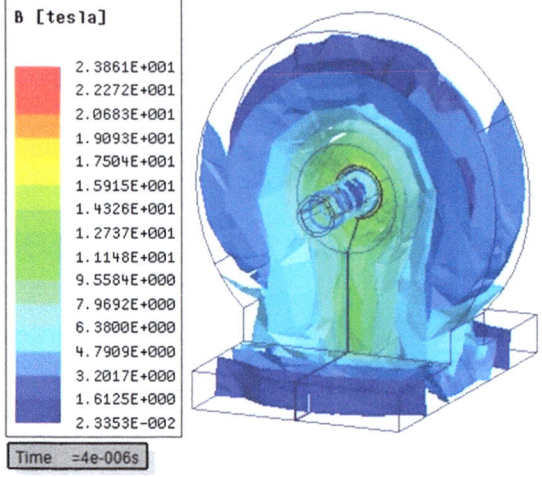

Fig. 9.6 Variation in surface force density with input voltages at multiple air gaps

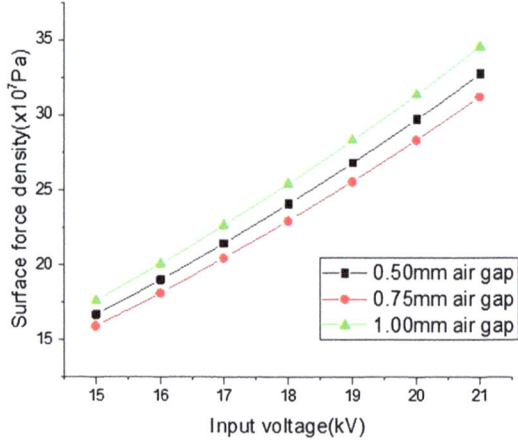

Fig. 9.7 Variation in magnetic field with input voltages at various air gaps

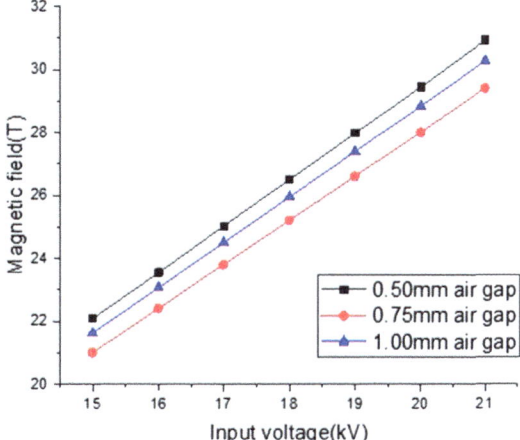

Fig. 9.8 Variation in input current with input voltages at various air gaps

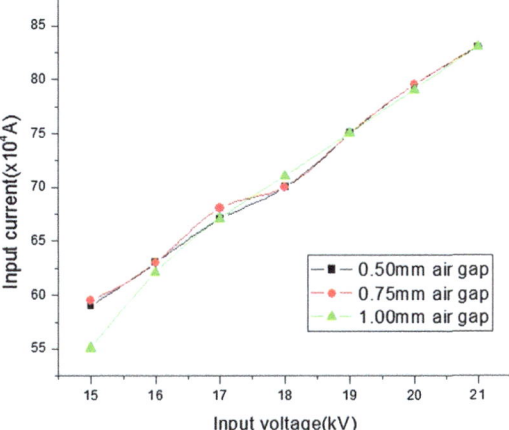

Table 9.5 Values of process parameters calculated by analytical model at various air gaps

Process parameter	Air gap (mm)		
	0.5	0.75	1.00
Magnetic pressure (MPa)	292.33	216.64	178.79
Magnetic field (T)	27.10	23.33	21.19
Input current (I)	711.25	612.28	556.20

9.4 Results and Discussions

To access the weldability of tubular arrangement discussed in Sect. 9.3, the weldability criteria were first estimated through analytical approach. The analytically calculated values shown in Table 9.5 provides minimum magnetic field and pressure

depict the variations of surface force density, magnetic field, and input current, respectively, with respect to change in air gap.

Surface force density analytically calculated for an air gap of 0.5, 0.75, and 1 mm is found to be 292.33, 216.64, and 178.79 MPa, respectively. In simulation, the input voltage at which all the three process parameters satisfy the analytically calculated values is referred as minimum required voltage for welding. Minimum required surface force density of 296.99, 229.51, and 200.68 MPa were obtained at 20, 18, and 16 kV, respectively. Again, maximum surface force density can never cross the UTS of material, so it restricts at 296.99, 283.34, and 282.99 MPa for 0.5, 0.75, and 1 mm air gap. Hence, the essential process parameter values required to identify the welding zone can be identified from Figs. 9.9, 9.10 and 9.11.

9.5 Conclusion

Simplified analytical calculation of the process parameters for EMW and their interrelation were described along with the coil design methodology. The present work renders a quick estimation of input voltage for various air gaps and proposes a weldability window considering surface force density, input current, and magnetic field simultaneously using 3D electromagnetic FE analysis. Simulation study for EMW of Al6061 and SS304 shows that for standoff distance of 0.5, 0.75, and 1 mm, the input voltage of 20, 18, and 16 kV is essential. The demonstrated methodology of developing a weldability window through finite element simulation would save cost and time to be spent in production of expensive setup and build material using MPW technique. Further, FE simulation can be conducted to predict the weldability window for different coil geometry and material combinations.

Acknowledgements The authors gratefully acknowledge the financial support provided to this study by the Advanced Technology Committee of BRNS Mumbai, India under project reference No. 2015013407RP00729-BRNS.

References

1. Findik, F.: Recent developments in explosive welding. Mat. Design **32**(3), 1081–1093 (2011)
2. Fukumoto, S., Tsubakino, H., Okita, K., Aritoshi, M., Tomita, T.: Friction welding process of 5052 aluminium alloy to 304 stainless steel. Mat. Sci. Technol. **15**(9), 1080–1086 (1999)
3. Park, Y.B.: Design of joints for the automotive spaceframe with electromagnetic forming and adhesive bonding. In: The Degree Doctor of Philosophy in the Graduate School of the Seoul National University, pp. 1–8 (2004)
4. Guo, X., Tao, J., Wang, W., Li, H., Wang, C.: Effects of the inner mould material on the aluminium–316L stainless steel explosive clad pipe. Mater. Des. **49**, 116–122 (2013)
5. Tabatabaee, M., Mahmoudi J.: Finite element simulation of explosive welding (2008)
6. Mousavi, A.A.A., Joodaki, G.: Explosive welding simulation of multilayer tubes. In: International Conference on Comput. Plast., Barcelona (2005)

7. Bahmani, A., Niayesh, K., Karimi, A.: 3D simulation of magnetic field distribution in electromagnetic forming systems with field-shaper. J. Mat. Process. Technol. **209**(5), 2295–2301 (2009)
8. Deng, J., Li, C., Zhao, Z., Tu, F., Haiping, Y.U.: Numerical simulation of magnetic flux and force in electromagnetic forming with attractive force. J. Mat. Process. Technol. **184**(1), 190–194 (2007)
9. Aizawa, T., Kashani, M., Okagawa, K.: Application of magnetic pulse welding for aluminum alloys and SPCC steel sheet joints. Weld. J. **5**(86), 119–124 (2007)
10. Kore, S.D., Date, P.P., Kulkarni, S.V., Kumar, S., Rani, D., Kulkarni, M.R., Desai, S.V., Rajawat, R.K., Nagesh, K.V., Chakravarty, D.P.: Application of electromagnetic impact technique for welding copper-to-stainless steel sheets. Int. J. Adv. Manuf. Technol. **54**(9), 949–955 (2011)
11. Kore, S.D., Date, P.P., Kulkarni, S.V., Kumar, S., Rani, D., Kulkarni, M.R., Desai, S.V., Rajawat, R.K., Nagesh, K.V., Chakravarty, D.P.: Electromagnetic impact welding of Al-to-Al–Li sheets. J. Manuf. Sci. Eng. **54**(9), 949–955 (2009)
12. Kore, S.D., Date, P.P., Kulkarni, S.V., Kumar, S., Rani, D., Kulkarni, M.R., Desai, S.V., Rajawat, R.K., Nagesh, K.V., Chakravarty, D.P.: Electromagnetic impact welding of copper-to-copper sheets. Int. J. Mat. Form. **2**(3), 117–121 (2010)
13. Raoelison, R.N., Buiron, N., Rachik, M., Haye, D., Franz, G., Habak, M.: Study of the elaboration of a practical weldability window in magnetic pulse welding. J. Mat. Process. Technol. **213**(8), 1348–1354 (2013)
14. Shim, J.Y., Kang, B.Y.: Distribution of electromagnetic force of square working coil for high-speed magnetic pulse welding using FEM. Mat. Sci. Appl. **4**, 856–862 (2013)
15. Xu, Z., Cui, J., Yu, H., Li, C.: Research on the impact velocity of magnetic impulse welding of pipe fitting. Mat. Design **49**, 736–745 (2013)
16. Cui, J., Sun, G., Xu, J., Xu, Z., Huang, X., Li, G.: A study on the critical wall thickness of the inner tube for magnetic pulse welding of tubular Al–Fe parts. J. Mat. Process. Technol. **227**, 138–146 (2016)
17. Kapil, A., Sharma, A.: Coupled electromagnetic–structural simulation of magnetic pulse welding. In: Advances in Material Forming and Joining, Springer India, pp. 255–272 (2015)
18. Nassiri, A., Campbell, C., Chini, G., Kinsey, B.: Analytical model and experimental validation of single turn, axi-symmetric coil for electromagnetic forming and welding. Proced. Manuf. **1**, 814–827 (2015)
19. Shanthala, K., Sreenivasa, T.N.: Review on electromagnetic welding of dissimilar materials. Front. Mech. Eng. **11**(4), 363–373 (2016)
20. Shribman, V.: Magnetic pulse welding of automotive HVAC parts. Rapport Tech. Pulsar Ltd **8**, 41–42 (2007)
21. Broeckhove, J., Willemsens, L.: Experimental research on magnetic pulse welding of dissimilar metals. Dissertation for the Master's Degree, Ghent University (2010)
22. Knight, D.W.: Solenoid inductance calculation, January (2013). http://g3ynh.info/zdocs/magnetics/Solenoids.pdf

Chapter 10
Electromagnetic Welding of Tubular Joints for Nuclear Applications

Surender Kumar Sharma and Archana Sharma

Abstract Electromagnetic Welding (EMW) technology is a promising and new manufacturing technology for welding of stainless steel alloys and aluminium alloys for nuclear applications. It has significant advantages over conventional welding techniques. A primary characteristic of this process is the use of non-contact electromagnetic forces to achieve welding of various metal workpieces. In this process, the welding is carried out by impact, when the workpieces are accelerated towards each other by the Lorentz force, produced due to magnetic field and the induced current in workpiece. The capacitor bank is required for generating high pulse discharge current at high frequency in the coil, which generates maximum magnetic pressure on the workpiece to obtain the weld. Electromagnetic Welding machines and weld coils are designed and developed for the welding of aluminium (Al 6061) and stainless steel (SS316L) alloys. This technique enables us to join similar and dissimilar metals, which are very difficult to weld by other conventional welding techniques.

Keywords Electromagnetic welding · Capacitor bank · Coil · Magnetic field

10.1 Introduction

Welding is the process of joining together two metallic workpieces so that bonding takes place at their boundary surfaces. In conventional welding processes, the welding is achieved by fusion, the workpieces are heated, melted and then cooled down so that the boundaries disappear and joint is created. Electromagnetic Welding (EMW) is a solid-state welding technology that uses electromagnetic

S. K. Sharma (✉) · A. Sharma
Pulsed Power and Electromagnetic Division, Bhabha Atomic Research Center,
Visakhapatnam, India
e-mail: surender80@gmail.com; surender@barc.gov.in

A. Sharma
e-mail: arsharma@barc.gov.in

© Springer Nature Singapore Pte Ltd. 2019
U. S. Dixit and R. G. Narayanan (eds.), *Strengthening and Joining by Plastic Deformation*, Lecture Notes on Multidisciplinary Industrial Engineering,
https://doi.org/10.1007/978-981-13-0378-4_10

forces to weld two workpieces together. The advantage of using EMW technology is that the formation of brittle intermediate phases is avoided so dissimilar metals can also be welded. The EMW technology is used for electrically conductive metals. EMW technology produces high-quality welds in similar and dissimilar metals in few microseconds' (<100 μs) duration. The workpiece is welded by fast-rising high pulsed magnetic fields produced due to pulse discharge current from the capacitor bank; this magnetic field induces an eddy current in the workpiece and rapidly accelerates the workpiece. The workpiece can be welded without any contact from a tool. It can be considered as one of the best high rate welding methods from the several aspects such as high cleanness, cost-efficiency and productivity. In the process, the workpiece can achieve velocities in the order of 100 m/s in less than 0.1 ms resulting in high-strain rates. Applying this technique, it is also possible to improve the weldability of some materials such as aluminium that is a good candidate material for use in automotive industry for lightweight structures and many other applications [1].

Nuclear applications are responsible for the development of EMW technology for the production of closing caps, end closures of nuclear fuel rods, metal canisters, stainless steel joints, aluminium container covers crimping and nuclear fuel pin. The need for a reliable method for joining high strength and low activation stainless steel alloys is central to the fabrication of fuel pins using these materials [2, 3]. EMW technology has been applied to limited applications in research laboratories such as welding of nuclear fuel rod [4], welding of similar as well as dissimilar metals [5] and metal to ceramic joints [6]. The results are encouraging but the technology did not reach industrial and nuclear applications due to limitations of the pulsed power driver. It is cumbersome, low efficient and large size and has high cost. Also, the coils are not fit for continuous operation and have low reliability. This technology has also some limitations and sometimes it is necessary to complement it with other methods. Solid-state pulse modulator for EMW applications offers an exciting opportunity to commercialize this technique in nuclear industries because of high reliability, high life, compact and high duty cycle [7].

10.2 Principle of EMW

EMW technology uses a high pulse current to generate high magnetic field and ultimately high pressure, which generate an impact force to bond and weld two metallic parts at an atomic level. The EMW system consists of a capacitor bank, switch, coil and the metallic workpieces to be welded. The capacitor bank is connected with a switch and then to the coil which is very close to the workpiece. When the switch is closed, the electrical energy stored in the capacitors is rapidly discharged through the coil as a high electric current pulse (>100 kA at >1 kHz) producing a pulsed magnetic field that induces opposite currents (intensity of the induced current is inversely proportional to the rise time of the pulse) on the nearby conductive workpiece (Lenz's law). The electromagnetic repulsion between the currents flowing in opposite directions and in close proximity provides the deformation force (Lorenz law) to the workpiece. The main process parameters that determine weld quality are as follows: current, frequency, magnetic pressure, stand-off distance, impact velocity and collision angle.

10.2.1 Current

The high frequency current in the coil produces an intense pulsed magnetic field that induces eddy currents in the workpiece. The magnetic field of induced current opposes the original change in the magnetic flux across the workpiece. The effect of this secondary current (induced eddy current) moving in the primary magnetic field in the coil generates Lorentz force, which accelerates the workpieces at very high velocities. This high-velocity impact causes plastic deformation on the workpieces, and under precisely controlled conditions (magnetic field, velocity and impact angle), an atomic level bond is created between the two workpieces. The intensity of the current determines the magnetic field and ultimately the magnetic pressures. The higher the current, higher will be the magnetic field and the magnetic pressure to accelerate the workpieces.

10.2.2 Frequency

Current and frequency are the most important parameters that must be computed, analysed and optimized for EMW technology; its effects on the EMW process are very tricky and complicated [8]. The important criterion for selecting the ringing frequency of the damped pulse discharge current can be obtained by equating the skin depth for the workpiece material with the workpiece thickness. The depth at which the induced current penetrates the workpiece depends on conductivity of the workpiece material and the frequency of the discharge pulse. High conductivity driver material such as copper or aluminium is used for welding materials having lesser conductivity.

10.2.3 Magnetic Pressure

Magnetic pressure accelerates the flyer metal at certain velocities, and then it impacts other metal to weld the pieces. The magnetic pressure generated due to induced eddy current will oppose the magnetic field from the coil and accelerates the flyer metal workpiece to gain velocity until it impact with other workpiece. The magnetic pressure must be high to achieve successful weld; otherwise, the flyer metal workpiece will impact into the other metal workpiece with lesser velocity and no joining will occur. High magnetic pressure can be obtained with high frequency and high pulse discharge current.

10.2.4 Stand-off Distance

Stand-off distance is the distance maintained between the two workpieces before the discharge of energy into the coil. This distance should be sufficient so that the flyer workpiece gets the required time to accelerate up to the maximum possible velocity at peak magnetic pressure points. But at the same time, it should not be too high that at the time of impact the velocity might have already started decreasing. Hence, the acceleration should continue to last till the impact of the workpieces. Keeping less stand-off distance causes the impact to take place before the flyer metal workpiece could reach the maximum velocity. For higher stand-off distance, the velocity drops to a lower value at the time of impact with other workpieces. So the stand-off distance should be optimized to achieve good quality weld.

10.2.5 Impact Velocity

The quality of the weld depends on the impact velocity between the workpieces during coalescence. The impact velocity accounts for the kinetic energy, which is then transformed into the energy used for the generation of bond between two workpieces. The process parameters that influence this impact velocity are the mechanical and the electrical properties of the materials, the electrical properties of the welding system, and the geometry of workpieces. The high impact velocity causes plastic deformation at the interface of the workpieces [9]. It also creates a plasma jet which removes any contaminants or oxide layers from both contact surfaces of the workpieces. A good quality weld is achieved when the impact velocity is greater than 150 m/s [10].

10.2.6 Collision Angle

High impact velocity and the optimized angle of impact are very important parameter to achieve a high-quality weld. Before the impact, the shock waves travel in both workpieces with a radial front and at a certain angle. The pressure peak is always at the impact point and at peak impact velocity. Successful and good quality welds are generally obtained when the impact velocity is between 150 and 1500 m/s and the impact angle is in the range of 5 and 20° [11, 12].

10.3   Theory of Electromagnetic Welding

The electrical energy stored in the capacitor bank is transferred into a coil through switch to generate high pulse current and ultimately high pulse magnetic flux density. The magnetic flux density (B) penetrates the flyer metal workpiece, and eddy currents with current density (J) are produced at the surface of the flyer metal workpiece. The Lorentz force ($J \times B$) accelerates the flyer metal workpiece until it collides with other metal. The equations governing the EMW processes are

$$\nabla \times J = -\sigma \left(\frac{\partial B}{\partial t} \right) \tag{10.1}$$

$$P = \left(B_{\mathrm{o}}^2 - B_{\mathrm{i}}^2 \right)/2\mu \tag{10.2}$$

$$B_{\mathrm{i}}^2 = B_{\mathrm{o}}^2 e^{\frac{-2t}{\delta}} \tag{10.3}$$

$$\delta = \frac{1}{\sqrt{\pi \sigma \mu f}} \tag{10.4}$$

where

σ is the electrical conductivity of the workpiece,
μ is the magnetic permeability of the workpiece,
δ is the skin depth,
B_{o} is the magnetic flux density at the outer surface,
B_{i} is the magnetic flux density at the inner surface,
f is the frequency of the pulse discharge current,
t is the thickness of the conductor metal,
P is the magnetic pressure and
J is the current density.

The circuit theory allows us to calculate the current, magnetic field and the pressure on the workpieces. The EMW process is equivalent to a primary R-L-C circuit coupled with secondary R-L circuit. The circuit diagram of EMW system is shown in Fig. 10.1.

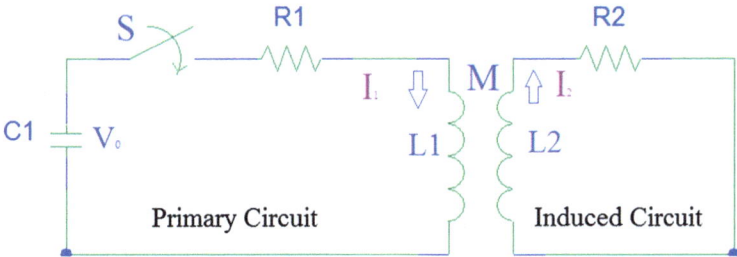

Fig. 10.1 Equivalent circuit of EMW system

In Fig. 10.1, V_0 is the initial voltage of the capacitor bank, C is capacitance of the capacitor bank, L is the inductance of the coil, R_1 is the resistance of the coil, M is the mutual inductance between the coil and workpiece, I_1 is the current in coil, L_2 is the inductance of the workpiece, R_2 is the resistance of the workpiece, M is the mutual inductance between the coil and workpiece and I_2 is the induced current in the workpiece.

When the switch (S) is closed, the capacitor bank discharges current (I_1) into the coil, the time-varying current in the primary circuit induces current (I_2) in the workpiece. The capacitor bank discharge current (I_1) and the induced current (I_2) in the workpiece can be obtained by solving the RLC circuit equations.

For the primary circuit,

$$R_1 I_1 + \frac{1}{C_1} \int I_1 dt + L_1 \frac{dI_1}{dt} + M \frac{dI_2}{dt} = 0 \tag{10.5}$$

and for the induced circuit,

$$R_2 I_2 + L_2 \frac{dI_2}{dt} + M \frac{dI_1}{dt} = 0 \tag{10.6}$$

By solving these equations, the pulse discharge current from the capacitor bank and the induced current in the workpiece can be obtained. The peak discharge current is given by

$$I_{max} = V_0 \sqrt{\frac{C_1}{L_1}} \tag{10.7}$$

The magnetic pressure (P) resulting due to the induced magnetic force (F) between the current in the coil and the induced current in the workpiece is given by

$$\frac{F}{l} = \frac{\mu_0 I_1 I_2}{2\pi d} \tag{10.8}$$

$$P = \frac{\mu_0 B^2}{2} \tag{10.9}$$

The total pressure (P) is the sum of two components [13]

$$P = P_d + P_a \tag{10.10}$$

where P_d is the pressure required for the deformation of the flyer tube, and P_a is the pressure required for the acceleration of the flyer tube.

The P_d can be calculated from the theory of thin-walled pressure vessel. It is given by

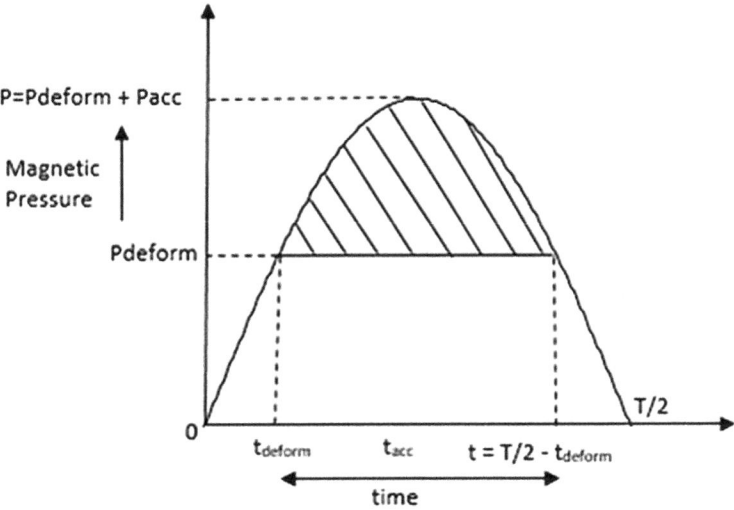

Fig. 10.2 The pulse magnetic pressure on a tubular workpiece with a coil

$$P_{\mathrm{d}} = \frac{2 \cdot t \cdot \sigma_y}{R} \tag{10.11}$$

where R is average accelerated tube radius, t is the flyer tube thickness and σ_y is the yield strength of the material.

The pressure (P_{a}) required to accelerate the tube, is calculated based on the assumptions that the movement of the tube is linear and that no other forces act on the tube and is given by

$$P_{\mathrm{a}} = \frac{t\rho V_c^2}{2s} \tag{10.12}$$

where ρ is the density of the tube material, V_c is the impact velocity and s is the stand-off distance.

The magnetic pressure (P) can be obtained from Eqs. (10.11) and (10.12). It rises initially and later falls after reaching the peak pressure (Fig. 10.2).

10.4   Simulations and Modelling for EMW

EMW technology can be identified as a coupled multi-physics problem, involving circuits, electromagnetic fields, impact and structural deformations. Mechanically, this is a high-strain rate process. The numerical simulations and modelling reduces the number of trial-and-error iterations required during the technological development and reduces the cost. The design stage can, therefore, be greatly helped through the support of computational models. The simulation of various circuits used in this process can be done using MATLAB and PSpice simulation software. Modelling of EMW of flat sheets is done by comparing the minimum impact velocities determined by ANSYS and ABAQUS software with the required minimum velocity obtained from analytical considerations [14]. EMW technology is a high-strain rate manufacturing processes with extremely high plastic strain regions. So, the Lagrangian analysis is not able to accurately model the process due to excessive element distortion [15]. Some models exist based on a coupling between finite elements and boundary elements for modelling the electromagnetic problem (ANSYS/Maxwell) and explicit approach for modelling the mechanical model (LSDYNA). The modelling of the material behaviour during high-strain rates is an important area to understand the EMW process. Various constitutive models have been proposed to take into account dynamic effects in material behaviour; it can be primarily sorted into three categories: Johnson–Cook models; the models derived from thermal activation analysis, and the models that are more specific to shock regimes and viscous drag.

10.5 Electromagnetic Welding Machine

The main components of EMW machine are capacitor bank, capacitor charging supply, energy transfer switch, welding coil and the workpiece. The capacitor bank is used for storing the electrical energy and then discharging the electrical energy to the welding coil. EMW machines are designed and developed for expansion and compression welding joints for the welding of Al and SS alloys.

10.5.1 40 kJ EMW Machine Using Single Sparkgap Switch

A 40 kJ EMW is designed with single sparkgap switch to carry out electromagnetic expansion welding of aluminium alloys. The EMW system has a capacitor bank, power supply, sparkgap switch trigger generator and the weld coil (Fig. 10.3). The capacitor bank has two numbers of 178 μF/15 kV/20 kJ capacitors connected in parallel, the triggered sparkgap switch is mounted on the top of the capacitors, and it can be pressurized by air/nitrogen up to 4 atm. The total inductance of the system was less than 200 nH and the short circuit discharge current and frequency of the system are 300 kA and 20 kHz, respectively. The capacitor bank is charged using a 15 kV/100 mA high voltage power supply. The capacitor bank and its power supply were isolated using a pneumatically operated decoupling switch after charging it to the rated voltage. The sparkgap switched is triggered, resulting in discharge of capacitor bank energy to the coil. The discharging circuit consists of a sparkgap switch, the welding coil and the aluminium tube-block workpiece. The

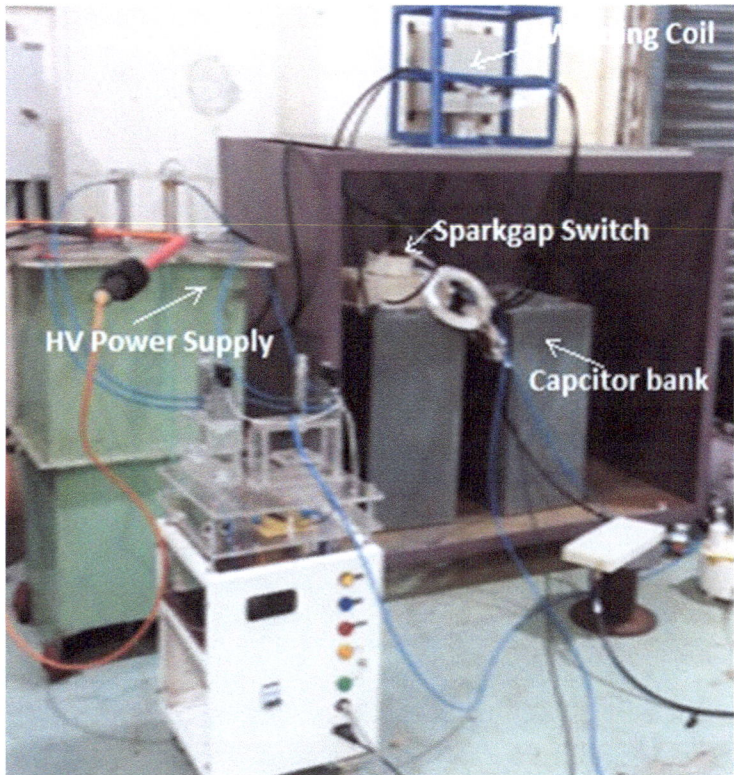

Fig. 10.3   40 kJ EMW machine

magnetic field developed around the coil induces eddy current on the inner aluminium tube that exerts magnetic pressure causing the inner tube to accelerate and impact on the outer flange. The system has delivered the maximum current of 170 kA, magnetic field of 20 T in the coil. The EMW system has been used for expansion welding of Al alloy tube to Al alloy block/flange.

10.5.2   180 kJ EMW Machine Using Multiple Sparkgap Switch

The 180 kJ EMW machine has 16 capacitors with stored energy of 11.25 kJ at the peak charging voltage of 75 kV. Each capacitor can deliver a peak current of 150 kA at peak charging voltage of 75 kV and the voltage reversal is limited to

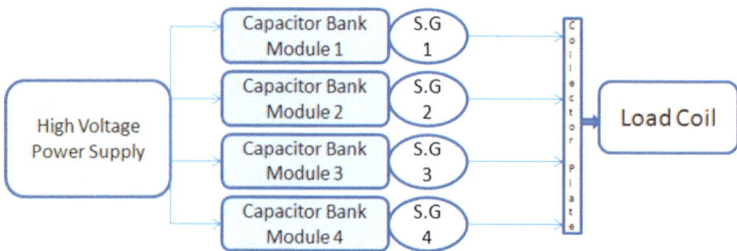

Fig. 10.4 Schematic diagram of capacitor bank with load coil

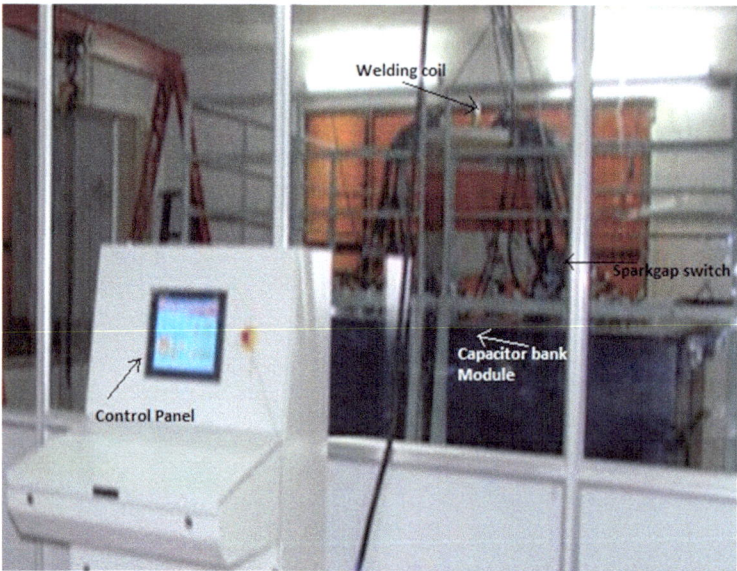

Fig. 10.5 180 kJ EMW machine

80%. The Equivalent Series Inductance (ESL) of each capacitor is 60 nH. The capacitor bank is subdivided into four modules with four numbers of capacitors in each module, connected by parallel plate transmission line and a single trigatron sparkgap switch. The high voltage output of all modules terminates to centralized collector plate. The schematic of implemented configuration of 180 kJ capacitor bank with load coil (Fig. 10.4) and the actual modular layout of capacitor bank have been shown in Fig. 10.5. The peak discharge current of 800 kA at 75 kHz is pumped in a single-turn coil for welding of SS alloy tube to SS alloy rod.

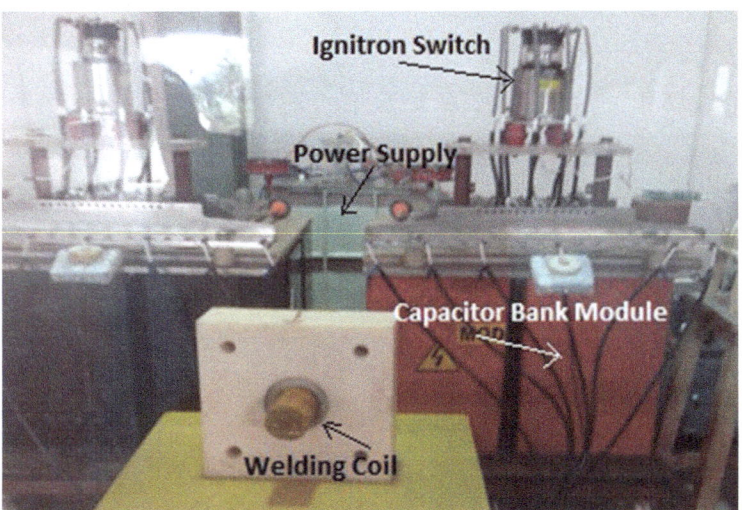

Fig. 10.6 400 kJ EMW machine

10.5.3 400 kJ EMW Machine Using Ignitron Switch

The 400 kJ EMW machine has eight capacitors with stored energy of 50 kJ each at peak charging voltage of 44 kV (Fig. 10.6). At the charging voltage of 44 kV, each capacitor can deliver a peak discharge current of 150 kA. The voltage reversal is limited to 10%. The ESL of the capacitor is less than 70 nH. The capacitor bank is subdivided into two modules with four capacitors in each module. The two ignitron switches are operated in parallel to transfer energy from the bank module to the load. The EMW system has pumped 200 kA current at 10 kHz and produced 22 T magnetic fields in the expansion welding coils.

10.5.4 800 kJ EMW Machine Using Multiple Railgap Switch

The 800 kJ EMW machine has 16 capacitors with stored energy of 50 kJ each at peak charging voltage of 44 kV (Fig. 10.7). There are four modules of capacitor bank and four numbers of capacitors are connected in parallel with parallel plate transmission line to each module and a railgap switch. The railgap switches are capable of transferring 10 coulombs (maximum) of charge at a peak current of 750 kA with a jitter of <2 ns. The outputs of all modules are terminated to centralized collector plate through coaxial cables. The 800 kJ EMW system has pumped 900 kA current at 15 kHz and produced 50 T magnetic field in the single-turn coil.

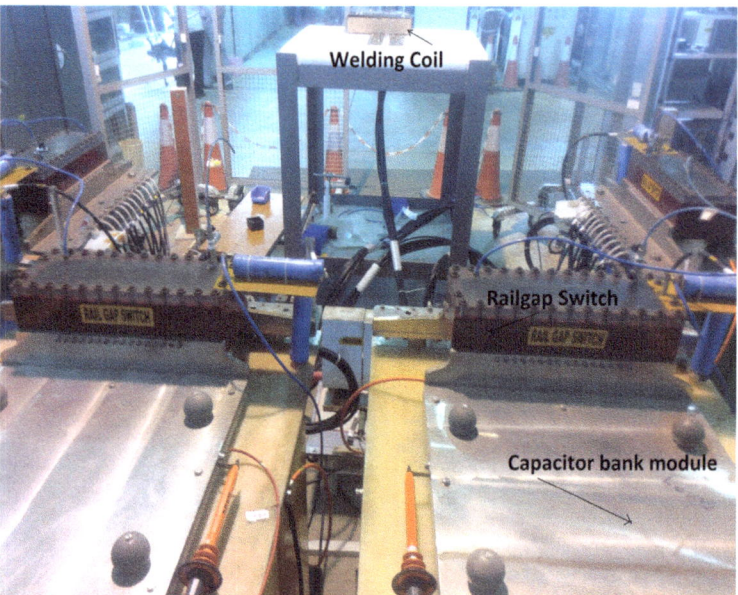

Fig. 10.7 800 kJ EMW machine

10.6 Electromagnetic Welding Coils

Coils are used to concentrate magnetic pressure into the electrically conductive workpieces placed in its vicinity. The welding coil consists of one or more electrical windings and is made from a highly conductive material. There are several factors to be considered while designing EMW coils such as material selection, operating voltage, magnetic flux and heating. Failure of the coils may occur due to high voltage breakdown, excessive heating or due to the magnetic pressure exerted on the workpiece. EMW technology requires development of stable coils for high magnetic field generation. The main characteristics of coil are high ultimate strength, high yield strength and high density. If magnetic pressure exceeds the material yield strength, residual plastic deformation takes place and accumulates in each successive pulse. This cause enlarges in inner dimensions and lower field so the coil must be replaced after few shots. Single-turn and multi-turn coils are mostly preferred for EMW of tubular size workpiece. The inductance of the welding coil can be calculated from the geometry and materials using inductance calculation techniques [16, 17].

10.6.1   Single-Turn Coil for Compression Welding

A single-turn coil is preferred because it has very low inductance which produces very high-frequency magnetic field across the workpiece. It has good mechanical, thermal and electrical isolation stabilities. Due to high frequency, the magnetic field is concentrated at the surface of the workpiece and expresses high magnetic pressure on it.

Magnetic field strength (*H*) is given by

$$H = NI/L \qquad (10.13)$$

where *N* (number of turns) = 1 as a single-turn coil was considered, *I* is the capacitor bank discharge current through the coil and *L* is the length of the coil.

Considering 50% coupling between the coil and workpiece, Magnetic field (*B*) is given by

$$B = 0.5\,\mu H \qquad (10.14)$$

A single-turn coil is made using Stainless Steel (SS) alloy SS-321 with a titanium insert (Fig. 10.8). SS321 stainless steel is Ni–Cr–Mo–type austenitic stainless steel, its performance is very similar to the SS304, but due to the addition of titanium, it has a better intergranular corrosion resistance and high-temperature strength. The inner diameter and inside length of the coil is 10.5 and 10 mm, respectively. The 3D model of the single-turn coil is shown in Fig. 10.9. These coils are used with 800 kJ EMW system at 15 kV to pump 900 kA current at 15 kHz and produced 50 T magnetic field.

Fig. 10.8 Single-turn coil with collector plate

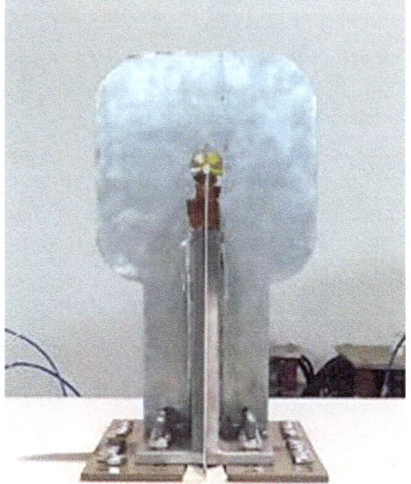

Fig. 10.9 3D model of
single-turn coil with titanium
insert

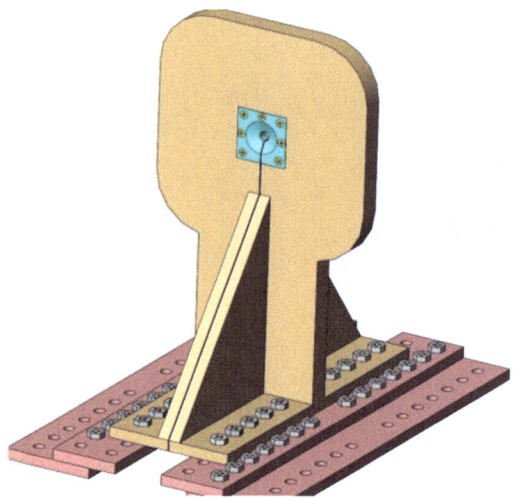

Magnetic fields of more than 100 T are generated and reported in single-turn destructive coils with an inner diameter up to 10 mm [18]. A single-turn destructive coil with triangular section is fabricated using 2 mm copper sheet (Fig. 10.10). The diameter and length of the coil were 10.5 and 10 mm, respectively. The coil was operated at 250 kA, and the magnetic field of 20 T is generated inside the coil.

10.6.2 Bitter Coil

Bitter coil is an electromagnet used for the generation of exceptionally strong magnetic fields. The upper bound of magnetic flux density is restricted by several factors. One principal restriction is the high stresses due to Lorentz forces in the coil. The Lorentz forces generate the distributed body force, which acts as the pressure of magnetic field. The common radial thickness profile of the Bitter coil is constant. A five-turn bitter coil is designed, fabricated and used for the compression welding of similar and dissimilar metals workpieces (Figs. 10.11 and 10.12). The inner diameter and the length of the bitter coil are 50 and 30 mm, respectively. The coil was operated at 200 kA and the magnetic field of 10 T is generated inside the coil.

10.6.3 Solenoid Coil for Expansion Welding

The solenoid coils were fabricated using copper conductor and high strength Fibre-Reinforced Plastic (FRP). FRP prevents the mechanical failure at higher

Fig. 10.10 Single-turn destructive coil

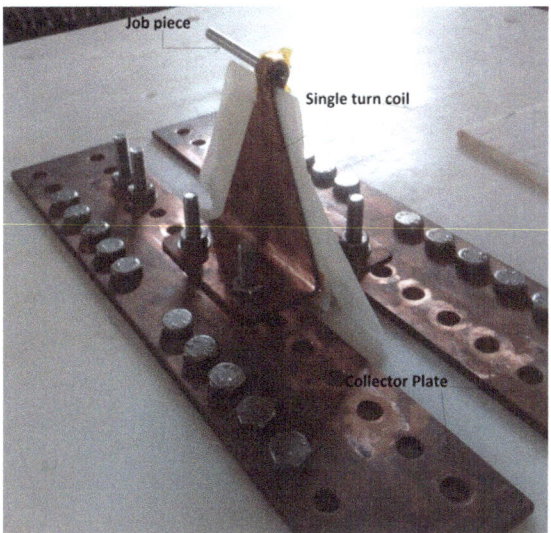

Fig. 10.11 Bitter coil (front view)

currents. The stresses that develop in a solenoid coil are the radial stresses, tangential stresses and the axial stresses. These stresses increase with the increase in current discharged into the coil and must be taken care of in order to prevent failure of the coil. A solenoid is designed, fabricated using copper wire with seven turns and FRP for expansion welding. The diameter and length of the coil are 64 and 40 mm, respectively (Fig. 10.13). These coils are repeatedly operated at 150–170 kA for EM expansion welding experiments.

Fig. 10.12 Bitter coil (side view)

Fig. 10.13 Solenoid coil for expansion welding

10.7   Expansion Welding of Aluminium Alloy Tube to Aluminium Alloy Block

Experiments are carried out to weld Al 6061 tube to Al 6061 block/flange. The electromagnetic expansion welding is influenced by different parameters like flyer tube thickness, stand-off distance, workpiece dimension and the system electrical parameters. Experimental studies were carried out on to obtain a good weld. The setup to weld tube to flange is made using Delrin block to maintain the concentricity (Fig. 10.14). This arrangement supported the tube-flange pair against the shocks generated due to the high discharge currents and pressures involved in the process.

The peak discharge current in the coil determines the magnetic pressure and the impact velocity. The magnetic pressure increases with the increase in the coil current as obtained from the simulation which ultimately increases the impact velocity (Fig. 10.15).

The deformation studies were carried out on 1.5 and 2 mm thick Al 6061 tubes of 65 mm diameter at variable discharge current provided by variable capacitor bank charging voltage. The diametrical expansion of 1.5 mm tube at 8–10 kV charging voltage, which results in 110–130 kA discharge current, is shown in Fig. 10.16. The diametrical expansion of 65 mm diameter tubes of 1.5 and 2 mm thickness is measured at different capacitor bank charging voltage (Fig. 10.17) and the coil discharge current (Fig. 10.18). The magnetic pressure on the workpiece with time at different charging voltages is shown in Fig. 10.19. It was observed that the thicker tube deforms less than that of thinner tube at the same energy level. It was found that a 65 mm diameter, 1.5 mm thick tube is expanded by 12.5% of its

Fig. 10.14 Al-6061 tube, Al 6061 flange and expansion welding coil

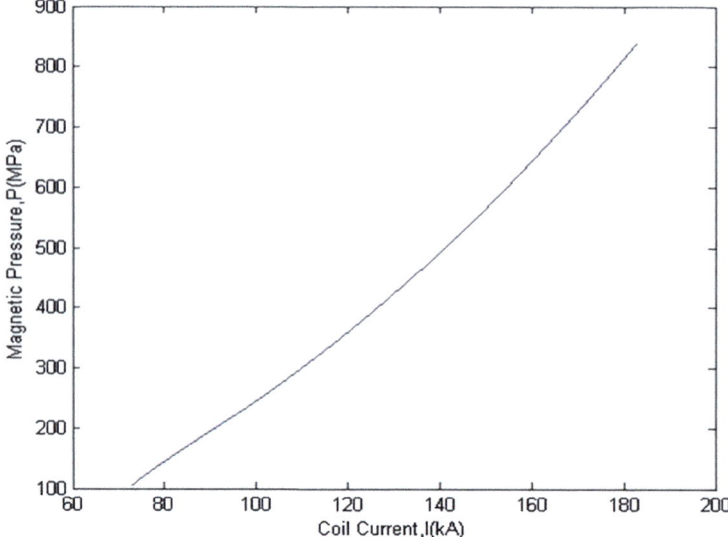

Fig. 10.15 Simulation of magnetic pressure with coil current

Fig. 10.16 Diametrical expansion of 1.5 mm Al 6061 tube at 8–10 kV

initial diameter whereas the 2 mm tube is expanded by 4.6% at 120 kA. It is also evident that the deformation of the tube increases with energy due to higher magnetic field and higher magnetic pressures generated at high discharge currents.

Finite element analysis simulation studies were done to validate the radial expansion data obtained during the experiment on a 1.5 mm thick Al 6061 tube of 65 mm diameter. It was found that the deformation (U) on the Al 6061 tube of

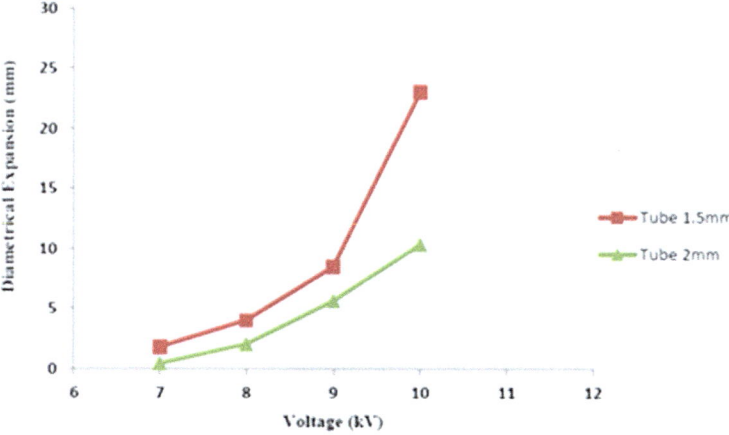

Fig. 10.17 Diametrical expansion of 1.5 and 2 mm tube at different voltages

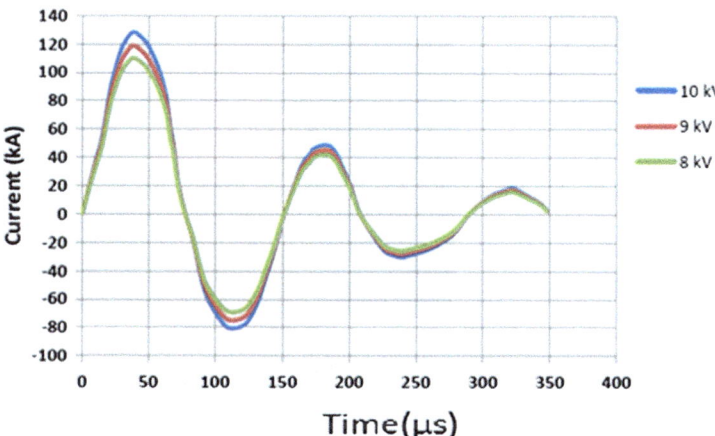

Fig. 10.18 Capacitor bank discharge current at different charging voltages

65 mm diameter in simulation at particular discharge energy was similar to the experimental data (Fig. 10.20). The deformation study is carried out using finite element analysis on 1.5 mm tube for the capacitor bank charging voltage of 9 kV and 120 kA discharge current. It generates a peak pressure of 180 MPa on the tube and a radial deformation of 4.01 mm is seen, which is similar to our experimental result.

Fig. 10.23 Copper driver on
the SS316l tube

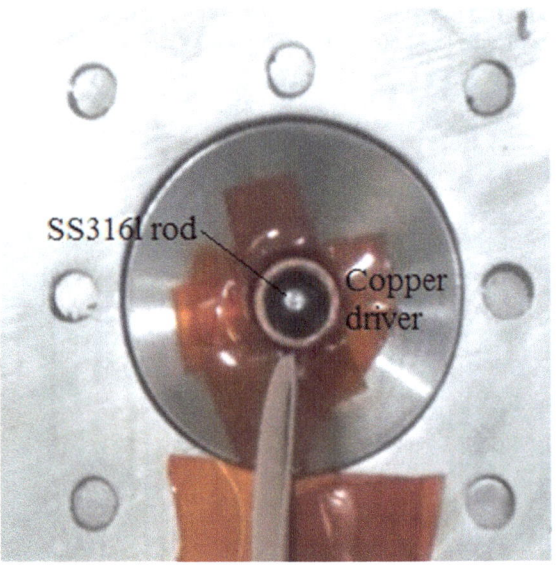

15 kHz. The workpiece was rigidly fixed on the Delrin holder (Fig. 10.22). The
thickness of the tube was less than the skin depth, so 0.6 mm thick copper driver
tube was used on the SS 316 l tube (Fig. 10.23). The pulse discharge current of
900 kA is pumped into the single-turn coil (Fig. 10.24), which generates 50 T field
inside the coil and it exerts >1600 MPa pressure on the workpiece. At this pressure,
the tube got welded to rod (Fig. 10.25).

The weld characterization of tube to rod was done using helium leak detection
and micrograph image. Leak rate of 10^{-7} torr.L/s is found and the wavy interface

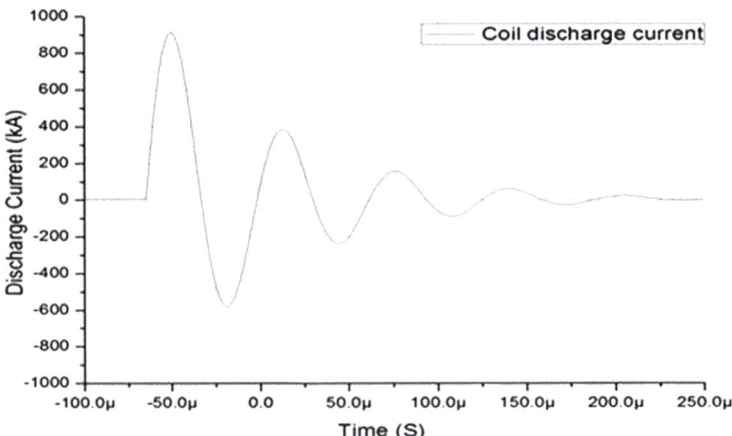

Fig. 10.24 Pulsed discharge current inside single-turn coil

Fig. 10.25 Welding of SS 316l tube to SS 316l plug

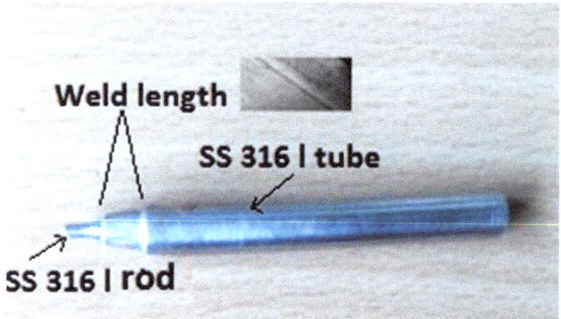

was seen on micrograph image (Fig. 10.26). The magnetic field and magnetic pressure at 900 kA peak pulse discharge current developed in the coil are calculated using Eqs. (10.9) and (10.14) (Fig. 10.27). The variable pressure acting on the coil due to pulse current is also calculated using Eq. (10.2) (Fig. 10.28). The workpiece was experiencing >1600 MPa pressure during welding, after 8–10 trials the insert piece got expanded which reduces the inductive coupling with driver material and also disturbs concentricity.

Dynamic explicit analysis of tube-rod workpiece joining is performed to calculate the stress, deformation and the impact velocity using finite element simulations. Boundary conditions and loading at one end of the tube is fixed and the other end is loaded with variable compressive pressure (Fig. 10.29). The Johnson–Cook constitutive model is used to simulate the material behaviour in impact of

Fig. 10.26 Micrograph image of SS316l tube to SS316l rod weld interface

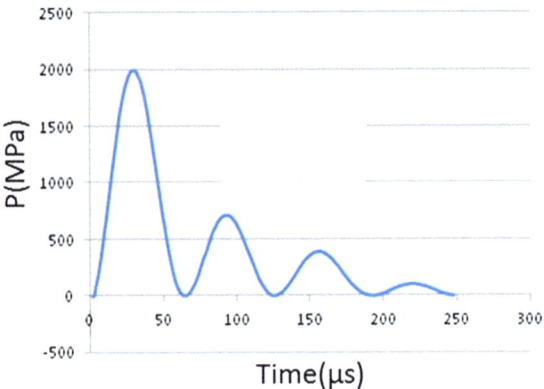

Fig. 10.27 Magnetic pressure on the coil

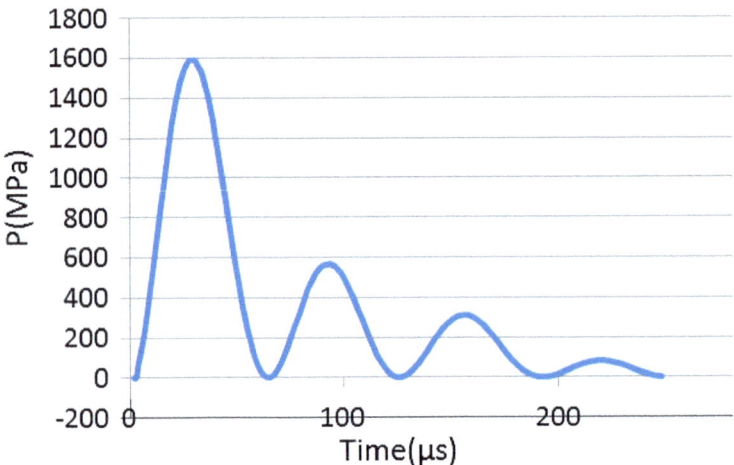

Fig. 10.28 Magnetic pressure on the workpiece

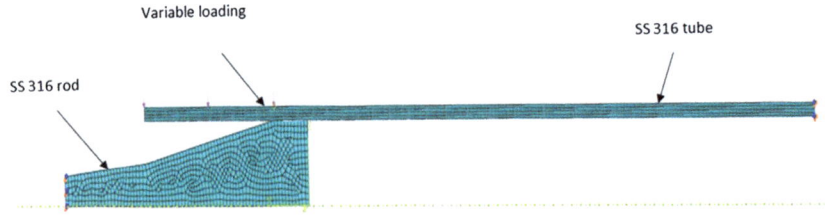

Fig. 10.29 Axisymmetric cross-sectional view of SS316L tube-rod workpieces

Table 10.1 SS316L parameters for Johnson–Cook model with standard strain rate form

S. No.	Parameters	Value
1	A (initial yield strength) MPa	280
2	B (strain hardening) MPa	1161
3	C (strain rate coefficient)	0.01
4	n (hardening exponent)	0.61
5	m (thermal softening exponent)	0.517
6	T_M (melting temperature) (°C)	1375
7	T_R (transition temperature) (°C)	25
8	ε'_0 (reference strain rate)	1
9	Specific heat J/kg K	500
10	Thermal conductivity (W/m K)	16.3

tubes. The three key material responses in the model are strain hardening, strain rate effects and thermal softening. These effects are combined in Johnson–Cook constitutive model:

$$\sigma_{\text{flow}} = [A + B\varepsilon^n]\left[1 + Cln\frac{\dot{\varepsilon}}{\dot{\epsilon_0}}\right]\left[1 - \left(\frac{T - T_R}{T_M - T_R}\right)^m\right]$$

σ_{flow} is the equivalent flow stress; A, B, C, n and m are the Johnson–Cook constants; T_R is the reference temperature at 300 K; and T_M is the temperature of melting (Table 10.1).

The stresses (S, in MPa) (Fig. 10.30) and the deformation (U, in mm) (Fig. 10.31) is calculated using finite element analysis at 900 kA peak discharge current during the time of impact. The displacement of the end point of tube with

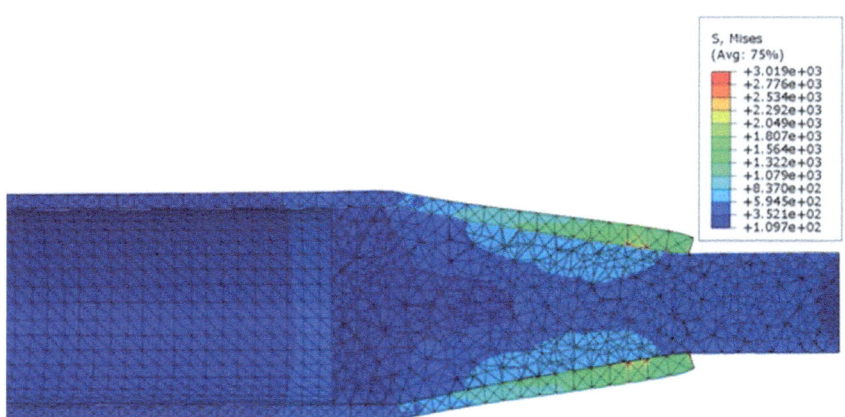

Fig. 10.30 Simulation of von Mises stress on the tube

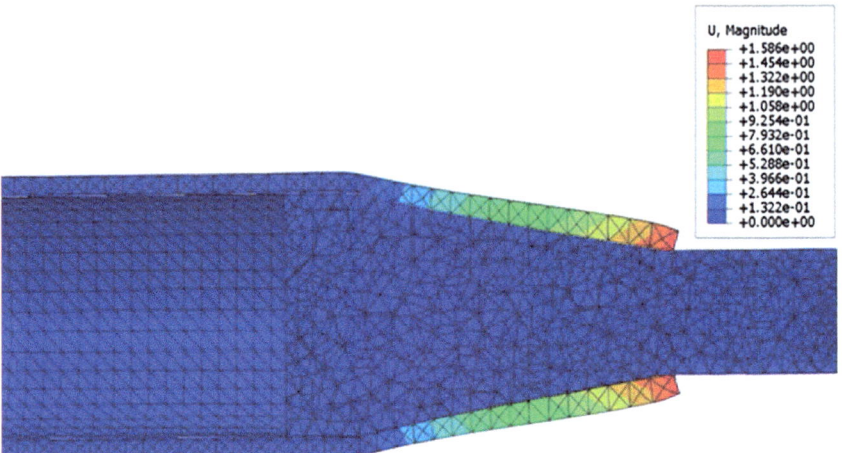

Fig. 10.31 Simulation of deformation at the end point of the tube

respect to time is simulated (Fig. 10.32), and the velocity of the end point is computed from the simulated data (Fig. 10.33). The impact velocity of SS 316L tube on SS 316L rod plug varies with time. The maximum impact achieved is 620 m/s at 7 μs.

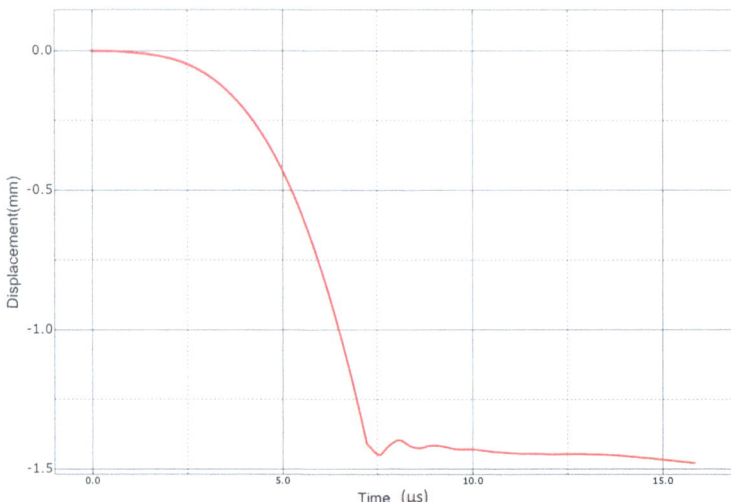

Fig. 10.32 Simulation of displacement of the end point with time

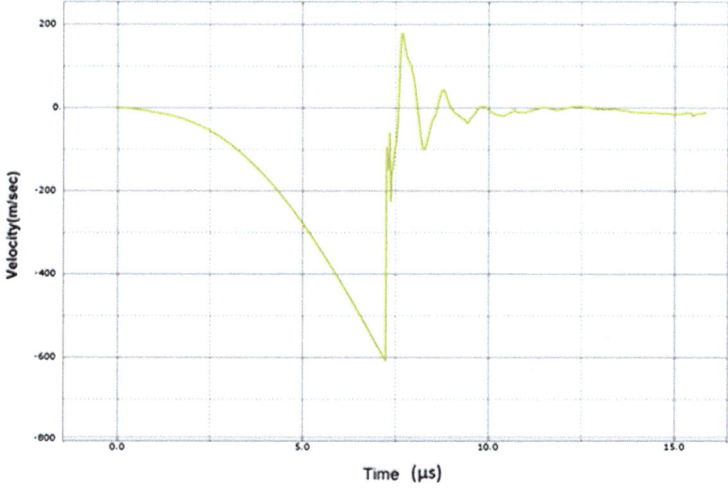

Fig. 10.33 Simulation of velocity at the edge of the tube with time

10.9   Electromagnetic Welding of Dissimilar Metals

Electromagnetic welding technique overcomes problems in welding of dissimilar material such as Cu–Al, Cu–SS, Al–SS sheets that could be encountered in the conventional welding techniques. The EMW technology is one of the reliable methods that can be used for the dissimilar metal joints. It proves to be a potential

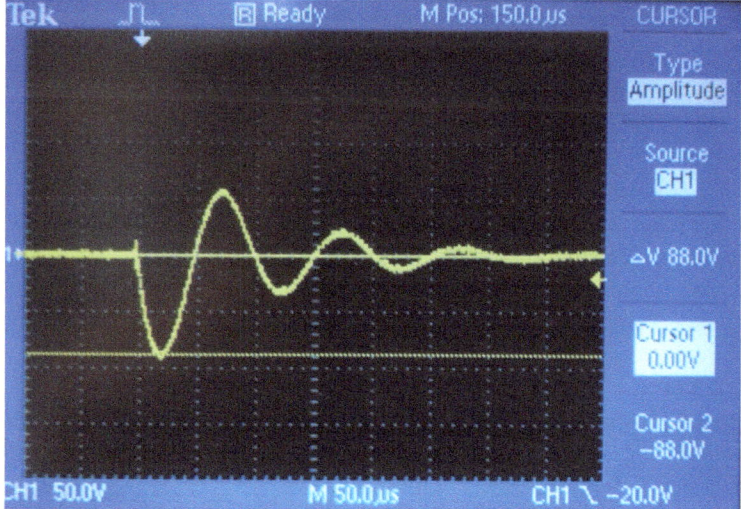

Fig. 10.34 Coil discharge current for copper tube to SS rod weld

Fig. 10.35 Welding of SS rod to copper tube

welding and manufacturing technology for various industrial components requiring dissimilar sheet metal and tubular joints. The influence of process parameters on the strength and metallographic structures of the electromagnetic welds can be studied in detail for different dissimilar metal combinations. Bitter coil was used for welding of copper tube with SS rod. Pulse current of 176 kA at 10 kHz is discharged into the bitter coil (Fig. 10.34). At this parameter, the welding of copper tube to SS rod was achieved and was confirmed by pull out, peel test and helium leak testing (Fig. 10.35).

10.10   Conclusion

EMW is a promising technology to weld similar and dissimilar metal joints at high productive rates. The welding is performed at very high speed and is achieved in <100 µs. This technology has many advantages over conventional techniques such as precision, reproducibility, high production rate, no tool marks, minimization of manual error, automation ease, etc., and the only limitation is that it can be used for conducting metal. EMW machines are designed and developed using high energy capacitor banks, high charge transfer switches and power supplies. Welding coils are also designed and developed for expansion and compression welding. Expansion welding of Al6061 tube to block and compression welding of SS316L tube are performed with these systems. Few aluminium alloys and SS alloys have excellent swelling resistance to neutron irradiation and irradiation creep behaviour properties and can be used for nuclear application. This technique overcomes the limitation for welding of special aluminium alloys and SS alloys for nuclear applications using conventional techniques.

Acknowledgements The authors wish to express thanks to Ms. Shobhna Mishra, Smt. Renu Rani, Shri J. M. V. V. S. Aravind, Shri Sukant Mishra, Shri Ekansh Mishra, Shri Pankaj Deb, Shri Ramanand Raman, Shri Nitin Waghmare and Dr. Rishi Verma of Pulsed Power and Electromagnetic Division and Shri Satendra Kumar, Shri M. R. Kulkarni, Shri P. C. Saroj and Shri Hitesh Choudhary of Accelerator and Pulsed Power Division for the development of EMW systems, for fruitful and useful discussions and experiments. The authors would also like to thank Shri R. K. Rajawat, Associate Director BTDG and Shri D. Venkateshwarlu, Regional Director, BARC Visakhapatnam for the motivation and support to carry out EMW research work. The author would also thank the organizers of AIMTDR-2016 to allow and present the research work in the conference.

References

1. Hameed, S., González Rojas, H.A., Sánchez Egea, A.J., et al.: Electroplastic cutting influence on power consumption during drilling process. Int. J. Adv. Manuf. Technol. **87**, 1835 (2016)
2. Brown, W.F., Bandas, J., Olson, N.T.: Pulsed magnetic welding of breeder reactor fuel pin end closures. Weld. J. **57**(6), 22–26 (1978)
3. Manoharan, P., Manogaran, A.P., Priem, D., et al.: State of the art of electromagnetic energy for welding and powder compaction. Weld World **57**, 867 (2013)
4. Kumar, R., Saroj, P.C., Kumar, S., Kulkrni, M.R., Kolge, T.S., Sharma, S.K., Shajju, A., Das, C., Sharma, A., Shyam, A., Bora, D.: Design and development of 100 KJ, 75 KHz electro-magnetic pulse welding system for ODS. In: IEEE Pulsed Power Conference, Texas Tech University, Austin, USA (2015)
5. Kore, S.D., Date, P.P., Kulkarni, S.V., Kumar, S., Rani, D., Kulkarni, M.R., Desai, S.V., Rajawat, R.K., Nagesh, K.V., Chakravarty, D.P.: Application of electromagnetic impact technique for welding copper-to-stainless steel sheets. Int. J. Adv. Manuf. Technol. **54**(9), 949–955 (2011)
6. Rajawat, R.K., Desai, S.V., Kulkarni, M.R., Rani, D., Nagesh, K.V., Sethi, R.C.: Electromagnetic forming—a technique with potential applications in accelerators. In: 3rd Asian Particle Accelerator Conference, p. 187, Gyeongju, Korea, 22–26 Mar 2004
7. Sharma, S.K., Shyam, A.: Development of compact rapid charging power supply for capacitive energy storage in pulsed power drivers. Rev. Sci. Instr. **86**(2), 023503 (2015)
8. Haiping, Y.U., Chunfeng, L.I.: Effects of current, frequency on electromagnetic tube compression. J. Mater. Process. Technol. **209**(2), 1053–1059 (2009)
9. Desai, S.V., Kumar, S., Satyamurthy, Chakravartty J.K., Chakravarthy, D.P.: Analysis of the effect of collision velocity in electromagnetic welding process of aluminium strips. Int. J. Electromag. Mechan. **34**(1), 131–139 (2010)
10. Loncke, K.: An exploratory study into the feasibility of magnetic pulse welding. Master thesis, Ghent University, pp. 1–147 (2009)
11. Zhang ,Y.: Investigation of magnetic pulse welding on lap joint of similar and dissimilar materials. The Ohio State University (2010)
12. Sartangi, P.F., Mousvi, S.A.A.A.: Experimental investigations on explosive cladding of cp-titanium/AISI 304 stainless steel. In: Lee, C., Lee, J.B., Park, D.H., Na, S.J. (eds.) Advanced Welding and Micro Joining/Packaging for the 21st Century, vol. 580–582, pp. 629–632. Trans Tech Publications Ltd., Stafa-Zurich (2008)
13. Lee, S.H., Lee, D.N.: Estimation of the magnetic pressure in the tube expansion by electromagnetic forming. Mater. Process. Technol. **57**(3), 311–315 (1996)

Printed by Printforce, the Netherlands